E Governance Data Center, Data Warehousing and Data Mining: Vision to Realities

INFORMATION SCIENCE AND TECHNOLOGY

Series Editor: Prof. KC Chen, National Taiwan University, Taipei, Taiwan

Information science and technology ushers 21st century into an Internet and multi-media era. Multimedia means the theory and application of filtering, coding, estimating, analyzing, detecting and recognizing, synthesizing, classifying, recording, and reproducing signals by digital and/or analog devices or techniques, while the scope of "signal" includes audio, video, speech, image, musical, multimedia, data/content, geophysical, sonar/radar, bio/medical, sensation, etc. Networking suggests transportation of such multimedia contents among nodes in communication and/or computer networks, to facilitate the ultimate Internet. Theory, technologies, protocols and standards, applications/services, practice and implementation of wired/wireless networking are all within the scope of this series. Based on network and communication science, we further extend the scope for 21st century life through the knowledge in robotics, machine learning, cognitive science, pattern recognition, quantum/biological/molecular computation and information processing, biology, ecology, social science and economics, user behaviors and interface, and applications to health and society advance.

- Communication/Computer Networking Technologies and Applications
- Queuing Theory, Optimization, Operation Research, Stochastic Processes, Information Theory, Statistics, and Applications
- Multimedia/Speech/Video Processing, Theory and Applications of Signal Processing
- Computation and Information Processing, Machine Intelligence, Cognitive Science, Decision, and Brian Science
- Network Science and Applications to Biology, Ecology, Social and Economic Science, and e-Commerce

For a list of other books in this series, visit www.riverpublishers.com

E Governance Data Center, Data Warehousing and Data Mining: Vision to Realities

Dr. Sonali Agarwal

Assistant Professor
Indian Institute of Information Technology (IIIT)
Allahabad
India

Dr. M. D. Tiwari

Director IIIT
Allahabad & Amethi
and
Former President
Association of Indian Universities
New Delhi
India

Dr. Iti Tiwari

Associate Professor
Uttar Pradesh Rajarshi Tandon Open University
Allahabad
India

River Publishers
Aalborg

Published, sold and distributed by:
River Publishers
PO box 1657
Algade 42
9000 Aalborg
Denmark
Tel.: +4536953197

www.riverpublishers.com

ISBN: 978-87-92982-72-8
©2013 River Publishers

Editors' Biography

Dr. Sonali Agarwal is working as an Assistant Professor in the Information Technology Division of Indian Institute of Information Technology (IIIT), Allahabad, India. She received her Ph. D. Degree at IIIT Allahabad and joined as faculty at IIIT Allahabad, where she is teaching since October 2009. She holds a Bachelor of Engineering (B.E.) in Electrical Engineering from Bhilai Institute of Technology, Bhilai, (C.G.) India and Masters of Engineering (M.E.) in Computer Science from Motilal Nehru National Institute of Technology (MNNIT), Allahabad, India. She worked as Lecturer and Assistant Professor at B.B.S. College of Engineering and Technology, Allahabad from 2002 to 2009. Her main research interests are in the areas of Data Mining, Data Warehousing, E Governance and Software Engineering. She has published widely in international journals and conferences. She has focused in the last few years on the research issues in Data Mining application especially in E Governance and Health. She has attended a number of National and International Conferences/workshops and she has to her credit more than 20 research papers in national / international journals and conferences. She has completed her Masters thesis work at Liverpool John Moores University (LJMU), Liverpool, U.K. during November 1999 to February 2000 under Indo-UK REC Project, a joint collaboration in between School of Computing & Mathematical Science, LJMU Liverpool UK and Motilal Nehru National Institute of Technology, Allahabad.

She has also taken part in Indo Swiss Joint Research Program (ISJRP) and full financial support was awarded to carry out joint research work and to gain knowledge regarding the recent research and experimental facility/work at EPFL, Switzerland, from December 2011 to January 2012. She is a Member of IEEE, ACM, CSI and Supervising two Ph.D. Scholars and several graduate and undergraduate students in the area of Data Mining, Data Warehousing, E Governance and Software Engineering.

 **Dr. Murli Dhar Tiwari** was born on August 16th, 1948 in a small village near Amethi in an ordinary family. After school education he completed his B.Sc., M.Sc. and D.Phil. Degrees from Allahabad University. In 1973 he joined HNB Garhwal University as a Lecturer, subsequently he became Reader, Head and Professor of Physics of the same University. In 1984 he joined University Grants Commission, New Delhi as a Principal Scientific Officer and after working for a decade on senior position there he then joined All India Council for Technical Education, New Delhi as a Senior Advisor in 1994.

In 1995 he joined MJP Rohilkhand University, Bareilly as a Vice Chancellor. During 3 years of his tenure he opened a number of Professional Courses at this University which was adjudged First University in all U.P. Universities in performance appraisal in three consecutive years. He also received Rs. 1.5 crores per year as an award to the University. After completing his term at Bareilly, Dr. Tiwari moved to Allahabad and established Indian Institute of Information Technology, Allahabad (IIIT-A) in 1999. Since then he isworking there as a Director.

The Institute runs B.Tech. (IT & E&C), M.Tech. in 7 areas, MBA and MS(CLIS).

The Institute got collaborations with most of the top class Universities of USA, Switzerland, U.K. and others. He has published about 150 research papers in refereed journals and supervised 12 Ph.D. students. He has enjoyed prestigious fellowships like Alexander-von-humboldt Fellowship and visited abroad several times.

Dr. Tiwari has also worked on prestigious positions such as President of Association of Indian Universities, New Delhi. At present he is Chairman, Electronics and Technology Division, Bureau of Indian Standards, New Delhi, Chairman, Indo-Swiss, Indo-Canada, Indo-Japan and Indo-Russian S&T Collaborative Programmes of Ministry of Science and Technology, Govt. of India.

To promote science education and research in the country, Dr. Tiwari organized A Conclave of Nobel Laureates in 2008, 2009, 2010, 2011 and the same is being done again in 2012. This is a unique programme and is contributing a lot for awareness and interests of young brilliant students for science education & research.

At present his interests are Information Technology in general-Wireless Sensors, Human Computer Interaction and Application of IT in untaped Non-conventional Energy in particular.

Dr. Iti Tiwari is Associate Professor in Uttar Pradesh Rajarshi Tandon Open University, Allahabad, India. Prior to that she has Reader & Head, Sociology Department at Jagadguru Rambhadracharya Handicapped University, Chitrakoot, Faculty in Rakshapal Bahadur Institute of Management, Bareilly and Coordinator (Distance Education), G. G. D. University, Bilaspur. Her specialization is in Child Crime and Handicapped. She has tremendous con- tributions in the area of physically challenged persons and Child crime. She has published more than 150 articles and manuscripts, 3 books and substantial research publications.

Contents

Table of Figures

Table of Tables

Abbreviations

1. **NEGP**: National E Governance Plan
2. **AI**: Artificial Intelligence
3. **RDBMS**: Relational Database Management System
4. **IR**: Information Retrieval
5. **VSM**: Vector Space Model
6. **OLAP**: On Line Analytical Processing
7. **DSS**: Decision Support Systems
8. **KM**: Knowledge Management
9. **CRM**: Customer Relationship Management
10. **OLTP**: On Line Transaction Processing
11. **CRISP-DM**: Cross Industry Standard Processes of Data Mining
12. **SQL**: Structured Query Language
13. **NN**: Neural Network
14. **SVM**: Support Vector Machine
15. **LS-SVM**: Least Square Support Vector Machine
16. **SVR**: Support Vector Regression
17. **SRM**: Structural Risk Minimization
18. **AAA**: Anytime, Anyhow, Anywhere
19. **G2G**: Government-to-Government
20. **G2C**: Government-to-Citizen
21. **G2B**: Government-to-Business
22. **DBMS**: Database Management System
23. **EDGI**: E Governance Development Index
24. **GPHRS**: Government Payroll and Human Resources System
25. **BOMNAF**: Border Management in Northern Afghanistan
26. **BOMCA**: Border Management Program in Central Asia
27. **ESL**: ESRI Lanka Initiative
28. **MMP**: Mission Mode Projects
29. **TIN**: Tax Information Network
30. **MCA**: Ministry of Corporate Affairs
31. **MEA**: Ministry of External Affairs

32. **CPO**: Central Passport Organization
33. **CRC**: Computerized Registration Centers
34. **CT**: Commercial Tax
35. **E&Y**: Ernst And Young
36. **NISG**: National Institute For Smart Government
37. **MIS**: Management Information Systems
38. **CCTNS**: Crime and Criminal Tracking Network and Systems
39. **CCEA**: Cabinet Committee on Economic Affairs
40. **DAC**: Department of Agriculture and Cooperation
41. **ICT**: Information and Communication Technologies
42. **PDS**: Public Distribution System
43. **CSC**: Common Services Centers
44. **PPP**: Public Private Partnership
45. **EDI**: Electronic Data Interchange
46. **NSDG**: National Service Delivery Gateway
47. **DCMI**: Dublin Core Meta-Data Initiative
48. **MDM**: Master Data Management
49. **GOI**: Government of India
50. **Deity**: Department of Electronics and Information Technology
51. **EGDMS**: E Governance Document Management System
52. **ODS**: Operational Data Store
53. **ETL**: Extraction, Transformation, and Loading
54. **PIP**: Performance Index Parameter
55. **OWFCM**: Optimized Weighted Fuzzy C Means Algorithm
56. **CLI**: Command-Line Interface
57. **GRE**: Gross Enrollment Ratio
58. **NER**: Net Enrollment Ratio
59. **AFP**: Acute Flaccid Paralysis
60. **PD**: Probability Of Default
61. **TIA**: Total Information Awareness
62. **HTF**: High Terrorist Factor
63. **KYC**: Know Your Customer

Preface

India is showing incredibly strong appearance in Information Technology (IT) sector worldwide, but the benefits of the revolution of Information Technology have not fully permeated into the day-to-day life of a common people, particularly in rural areas. National E Governance Plan, initiated by the Government of India, first time under a concerted effort is being made to take Information Technology to the masses in areas of concern to the common man. It aims to make most services available online, ensuring that all citizens would have access to them, thereby improving the quality of basic governance on an unprecedented scale.

Since all government organizations are now executing their tasks on the basis of digital information, the consequences of inadequate processing power, storage, network accessibility, or data availability can have a profound impact on the viability of the organization itself. A Data Center establishment for E Governance could help to alleviate some of these issues, successfully integrate new solutions that can increase productivity and lower ongoing operational costs.

Data Warehousing and Data Mining are the emerging technologies which have experienced significant enhancement in the last decades and achieved a promising level of advancement. These techniques are broadly applied today in several fields of commercial, industrial as well as scientific research. In the present scenario Data Mining, Decision Support, Pattern Discovery and Trend analysis have a greater impact on businesses and engineering alike. Given this evolution, it is important to understand the potential advantages of Data Mining and Data Warehousing and their positive effects on E Governance applications.

This book brings together a set of emerging E Governance applications areas which could be benefited through evolving application domains of Data Warehousing and Data Mining. It is believed that this book assists in evolution of E Governance application ideas, highlighting collaborative need, exchange of good governance practices, developing a common strategy, and promoting interoperability which would allow large savings in costs and provide citizens a seamless view of the Government.

This book also attempts to disseminate information on about several E Governance projects and possible Data Mining benefits which are the prominent future of better governance in India. Strategic Management of these projects through Data Mining would certainly encourage policy makers to understand better models of E Governance, thorough evaluation of projects, tracking of the objectives and to develop a more enhanced collaborative design towards completion and execution of the National E Governance Plan. This revolutionary approach will help the government to evaluate the mandatory requirements and develop competencies for upcoming project management, thereby improving the lives of billions. It will also help innovators to think for unique solutions to achieve the end objectives of E Governance.

It would be useful for regular students, application developers, government officials, policy makers, as well as researchers involved in E Governance and Data Mining Applications. This book presents a complete portrait of the E Governance Data Mining applications, including Data Management Framework, Data Center and Data Warehousing applications along with possible research directions. An imperative inspiration for writing this specific book was the need to build an inter-organizational framework for E Governance using Data Mining—a challenging task due to the extensive and multidisciplinary characteristics of this fast-developing research field. We mightily hope that this book will encourage the people with different scientific backgrounds and have experience to exchange their precious views regarding the Applications of Data Mining in E Governance so as to contribute toward the establishment of good governance for the developed society.

Dr. Sonali Agarwal
Dr. M.D. Tiwari
Dr. Iti Tiwari

1

Roadmap To E Governance Data Management, Data Center, Data Warehousing and Data Mining

1.1 Introduction

Application of internet, information systems, computers and communication technologies in developmental, regulatory, social welfare department is gaining momentum in world. Out of more than five billion population, approximately 78 percent of the worldwide population is well connected but the computer and communication technologies awareness in least developed countries still requires substantial stimulation for developmental, regulatory and social welfare activities keeping the population and diversity in view. It is also established that the whole world is now behaving enthusiastically for sharing of E Governance plans, policies and experiences but diversity in language, religion, climate, health conditions, and extent of facilities enjoyed by each country differs to the great extent, which leads to a very complex society worldwide.

The book extensively pays attention to the area mainly related to E Governance, Data Management, Data Center, Data Warehousing and Data Mining. E Governance is certainly a revolution in the government services, to provide full availability, accessibility, efficiency with transparency and accountability to the whole society. The most essential requirement of E Governance projects are computerization of different government organizations. Increasing trend of computerization in different government departments such as regulatory, developmental including social welfare involve a large data. Subsequently a universal framework is required to achieve efficient data management including data storage and knowledge sharing in Local, State and Central Government. Data Mining is a well-organized way of discovering association from huge databases for Knowledge Management and to develop strategy for

E Governance Data Center, Data Warehousing and Data Mining: Vision to Realities, 1–10.

futuristic planning and subsequently implementation of task within deadlines. Knowledge Discovery in databases through Data Warehousing and Data Mining includes techniques to identify, characterize and allocate knowledge for learning and awareness specially when huge data generation are involved at the different levels of governments and appropriate decision making is imperative for efficient resource management.

1.2 E Governance

E Governance, an emerging practice, includes various processes, which explore the potentialities of Information and Communication Technology applications. The E Governance projects are aspired to assimilate government organizations at different levels and as well as public and private sectors. These kinds of integration of different government / non-government working organizations are extremely helpful to establish good governance in developing countries having inadequate resources with surplus manpower and deficit finance.

The prime responsibility of E Governance is to facilitate various citizen centric services starting from birth of person such as Health and Education till the death of a person such as Pension and gratuity [1.1], which is still a challenging task due to digital divide and over population. Implementation of various citizen centric services such as developmental, regulatory and Social Welfare services could be accelerated through strategic management based on appropriate Data, Data Center and Data Warehouse with Data Mining tools. The book extensively pays attention to the area mainly related to E Governance, Data Center, Data Mining and Warehousing.

E Governance is certainly a revolution in the government services, to provide full availability, accessibility, efficiency with transparency and accountability to the whole society. The achievements of E Governance are quicker assessment, reduction of replication of work, effective and full utilization of financial and human resources, knowledge gained and crisis management. The most essential requirement of E Governance projects are computerization of different government organizations [1.2]. Increasing trend of computerization in different government departments such as regulatory, developmental including social welfare involve large data. Subsequently a newly developed framework is of the essence which facilitates efficient data storage and knowledge sharing in Local, State and Central Government.

Although, all the developing countries are equivalent to developed countries in terms of policy development but are generally poor implementer

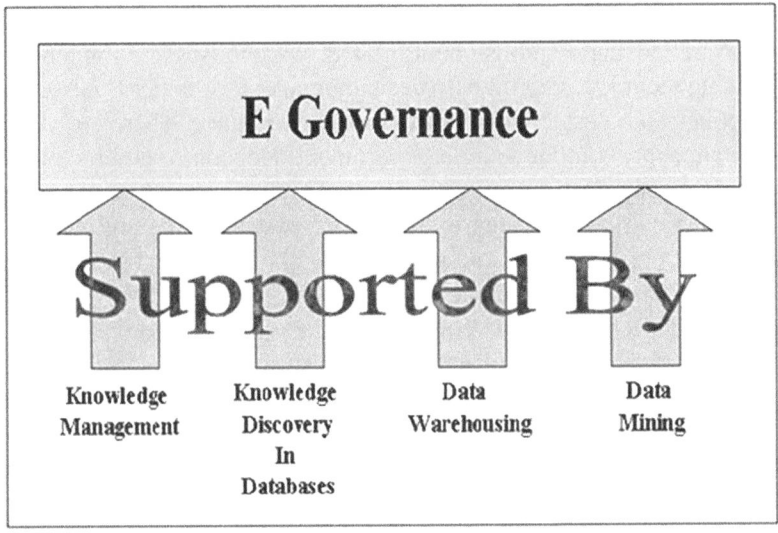

Figure 1.1 E Governance Enabling Technologies

caused by resource constraints and delays in implementation leading to time overrun and consequently financial overrun [1.3]. E Governance system could be quickly accessed by using efficient Knowledge Management techniques. It supports storage, retrieval and distribution of information associated to government management and facilitates useful interaction and instantaneous dialogue with all associated guidelines dealing with all stakeholders including public in India. The Figure 1.1 illustrates four pillars of E Governance.

The E Governance projects started with many initiatives at different level of computerization and information systems. Existing information systems are associated with limited storage capacity for operational data. Efficiency of E Governance could be only judged because of accurate information and quicker citizen services. But conventional databases are not adequate to realize the requirements of E Governance. There are some issues, indicated below:

- Which people are the least expected to default on their tax payments within a given dataset?
- What is the effect on Revenue generation if Department of Railways raises the fare of the trains?
- What would be the net outcome on over all skill development if Department of Education emphasizes easy course contents of the various courses as opposed to its technical capabilities?

Above issues are also investigated at appropriate level through Data Mining and Warehousing. Establishment of Data Warehouses and implementing Data Mining techniques could help decision makers for better E Governance service policies and establishment of Good Governance. There are varieties of important aspects influencing a government decision to embrace E Governance as an approach to governmental transformation. Sometimes, government officials may be trying to break with past practices and widen new processes and organizational cultures to bring about a foremost alteration in how E Governance is implemented.

In additional cases the government officials may be motivated by consideration in transforming and improving existing functions to attain better competence and improved service delivery. Instant transfer of data to the review authority will ensure no change or manipulation in the data and obviously such data will reflect the reality of the circumstances. This will also lead to full control of government at state level or Nation level. The numbers of motivations are myriad, but the five basic requirements: availability, accessibility, efficiency, transparency and accountability are essential for successful completion of any E Governance Project. A few of the most considerable reasons used to justify E Governance initiatives are shown in Figure 1.2.

1.2.1 Availability

The availability of relevant information about E Governance projects to citizen at large, government officers, business community and all other organizations could be ensured through the presentation on their websites. This could be strictly implemented and uniformly available for all the citizens. All the deadlines for compliances of regulatory measures should also be available on their websites.

1.2.2 Accessibility

The fast acceptance of Internet and Information Technology has made easier and cheaper access to information. All useful information and services are carried with the assistance of websites to serve people at all the time. Accessibility of government services through electronic media also promotes efficiency and transparency [1.4]. Worldwide accessibility is essential and it requires physical, environmental and technical support. Another inspiring interest of E Governance schemes are promotion of citizen participation in government related activities. Active participation may guide citizens to experience a significant role in the government execution, which possibly will

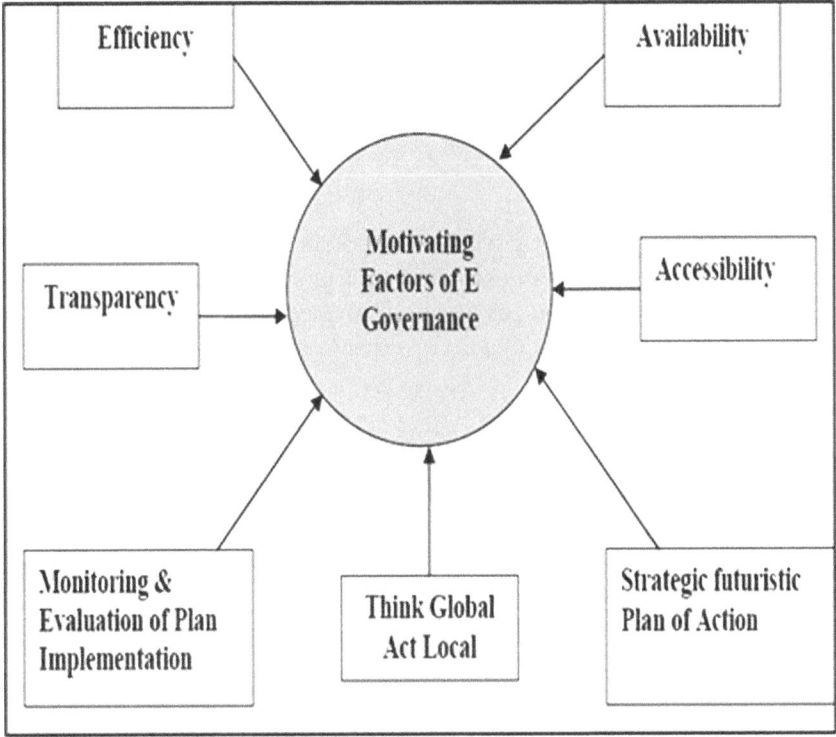

Figure 1.2 Motivating Factors of E Governance

help to construct enthusiasm for the promotion of E Governance projects even further.

1.2.3 Efficiency

The most valuable advantages of using E Governance projects are efficiency, which is an economic term for circumstances that generate the largest profit with smallest costs. In E Governance projects, efficiency may be achieved by putting the following efforts.

- Automation of the standard task: It may diminish errors of the projects and get better consistency in results.
- Re-engineering and reformation of working measures: It may help in cost reduction and finally improve efficiency.

- Reduction of the repetitive tasks: It is helpful to improve efficiency. An efficient working environment develops new skills in employees. The efficiency of both projects in terms of execution must be available for citizen information.

1.2.4 Accountability

Accountability suggests that government system is committed to carry out services according to certain rules or protocols, under a fixed deadline and with a predetermined use of resources and performance standards. Accountability is essential to build confidence in E Governance system.

1.2.5 Transparency

As government bearing large expenditure for computerization of services in national, international markets, it is very important that proper selection criteria must be espoused to make sure that government deals be awarded to the most competent parties. Transparency is visibility and clarity of laws, regulations, and procedures. Transparency has following dimensions [1.5].

- All Government decisions, verdicts and resolutions.
- All Government services.
- Every Government transaction details.
- Every Action of different Government Departments.
- Ongoing and completed projects.

1.2.6 Think Global Act Local

The world is turning into extremely large marketplace. The strategy should be based on "Think Global Act Local" during implementation of any project. E Governance provides the opportunity for small scale projects to be planned globally and launched locally with less resource and in local languages [1.6].

The present investigation aspires at developing a suitable E Governance model framework based on Data Mining and Data Warehousing techniques which may be efficiently used by the government at all its administrative levels National/State/District/Block).The proposed Model will serve all possible aspects of E Governance with the help of four basic building blocks:

- Administrative Block
- Technical Knowhow Block
- Service Block
- Stakeholder Block

1.3   Need of Data Management in E Governance

Any Government system produces massive quantity of data as its core organizational asset through its various departments. In E Governance, Data management is essential to develop, implement and control various government activities and policies for efficient utilization of all its resources. It covers the complete life cycle of data including all issues related to data creation, storage, updation, retention and delivery. It needs common data standards and practices for Data Management so that the coordination among heterogeneous data sources and multiple categories of users could be possible [1.7].

1.4   E Governance Data Centers

E Governance Data Centers are one of the key components of the E Governance model framework as proposed in National E Governance Plan (NeGP) [1.8]. E governance Data centers may be established at country level, state level and local administration level through high speed connectivity and common point of access. The primary goals of E Governance Data Center are single window citizen service, any time any where government access and secure data storage for efficient E governance services. This also ensures efficient, quicker and timely delivery of citizen centric services.

1.5   Data Warehousing for E Governance

The success of an E Governance program is mainly depending upon citizen satisfaction which could be easily achieved by timely and quicker citizen centric services [1.9]. To achieve this, digitization of various governmental data have been already started and the size of data is increasing enormously. Although, data is available, still it is difficult to achieve timely and quicker citizen centric services. It needs efficient user friendly and intelligent decision making tools which less technical expertise. An E Governance Data Warehouse would greatly facilitate efficient access of E governance data for all its stakeholders to achieve meaning full information at correct time.

1.6   Application of Data Mining in E Governance

Data Mining is a well-organized way of discovering association from huge databases for Knowledge Management and to develop strategy for futuristic planning and subsequently implementation techniques accomplish the task within deadlines. Knowledge Management includes techniques to identify, characterize and allocate knowledge for learning and awareness specially when huge data generation are involved at the entire levels of governments and appropriate decision making is imperative for efficient resource management.

In any government system decisions are always based on previous practices and experiences. Once Data Mining system has been adopted, conventional practices have been improved and best practices could be introduced to complete the task in time. Moreover Data Mining could facilitate quick and useful knowledge extraction from huge data sources at reduced costs and finally ever-increasing organization growth and development opportunities [1.10].

1.7   Focus of this Book

The end objective of this book is to manifest and assert data, data warehousing and Data Mining strength in E Governance application. Data is a valuable asset in E Governance therefore an E governance Data management framework is presented with its all policies and procedures. Data Center is an integral component of an E Governance infrastructure which could be helpful to facilitates single window government for all kind of citizen services. A Data Warehouse is a systematic storage of data collected from various government sources over a period of time. Data Mining provides efficient techniques for government agencies to analyze data quickly at lower costs. The data extraction process generates interesting hidden patterns. The revealed unseen patterns enable the government systems in making better decisions and advanced plan for serving the citizens. The objective is to prove the efficacy of Data Warehousing and Data Mining in E Governance and develop a model which facilitates the following:

- Efficient Data Management techniques for E Governance.
- Efficient Data Center implementation framework especially for E Governance Applications.
- Efficient Data Warehouse model framework for National, State and District level E Governance Applications.
- Efficient Data Mining techniques including Classification, Clustering and Regression for improved internal processes and government policies.

1.8 Outline of the Book

Chapter 1, the introductory chapter includes the statements which propelled us for adopting such area of research. More specifically, it contains a Roadmap to E Governance Data Management, Data Center, Data Warehousing and Data Mining. It highlights the necessity of Data Management, Data Centers, Data Warehousing and Data Mining in E Governance.

Chapter 2 of this book perceives a detailed Literature Review of various Knowledge Management, Data Management, Data Center, Data Warehousing and Data Mining application in E Governance. This chapter also dissertated about the E Governance using Data Mining perspective.

Chapter 3 demonstrates the world wide status of E Governance. In this chapter status of E Governance is presented with the help of five geographical divisions such as Africa, America, Asia, Europe and Oceania.

Chapter 4 consist status of E Governance in India. It includes the significance of Mission Mode Projects. It also discussed about various mission mode projects undertaken by Central government, State government and those projects which are considers under both Central and State Government bodies.

Chapter 5 covers the conceptual framework on data management for E Governance. It emphasizes on existence of data as an organizational asset which needs a defined data management policy framework for creating a strong foundation of E Governance model. It includes various issues related to Data Management, Data Architecture, analysis and design. A detailed description of Data Security, Data Quality and Metadata management has been presentation here in this chapter.

Chapter 6 presents the detailed description of the need of Data Center in E Governance applications. This chapter elaborates the desirable features of E Governance Data Center. The Data Center will facilitate a central repository of data for efficient and reliable access. It could be developed at different level administrative units for better monitoring and controlling of E Governance services. It also highlights the Importance of Cloud Computing in Data Centers.

Chapter 7 demonstrates a Data Warehouse model for E Governance. The chapter evaluates the importance of present E Governance system and proposes a suitable Data Warehouse based E Governance Model Framework along with Data Mining application for efficient transparent, accountable and effective E Governance in the light of Information Technology policy.

Chapter 8 is based on two case studies i.e. E Governance in Education and E Governance in health. In this chapter two real time projects i.e. "Education for All" and "Pulse Polio Immunization" has been considered and the significance of clustering classification and regression approach as a strategic management tool for E Governance practices has been presented.

Chapter 9 demonstrates the Data Mining applications in different departments including Food and General Supplies, Health, Rural Development, Banking, Agriculture, Planning, Commerce and trade and judiciary and also in the State Government activities. It highlights various ways of quicker information retrieval, development of decision support systems, fraud detection and preparing future policies based on previous experiences.

Chapter 10 consist the recent areas of E Governance where Data Mining and Data warehousing has been implemented with significant impact. In this chapter the Data Mining and Data Warehousing applications in Medical field, chemical industries, telecommunication industries, banking industries, e-learning and fraud detection has been discussed as per real time trends. Chapter 10 also covers the recent developments such as opinion mining, web usage mining and text mining. It also includes the possible framework of E Governance system which could avail the benefits of opinion mining, web usage mining and text mining.

Chapter 11 summarizes all the observations, analysis of the established facts and interpretation of the obtained research results on the basis of SWOT analysis and PESTLE analysis. This chapter is helpful to derive the efficacy of the anticipated model of E Governance, through statistical, analytical review and keeping focus on issues of implementation. Chapter 11 also reviews the recent research work and suggests future perspective in the area Data management, Data Center, Data Warehousing and Data Mining Based E Governance applications.

2

Foundation of Data, Data Warehousing, Data Mining and E Governance

2.1 Introduction

This chapter includes detailed study about Data, Data Warehousing and Data Mining for E Governance in chronological order. The first part concerns the history of Database developments, Knowledge discovery techniques, Data Warehousing and Data Mining. The second part draws attention towards origin of Knowledge Management and efficient use of Data Mining techniques as decision making processes. The third part discusses about the basics of E Governance and finally conclusions have been discussed.

2.2 History of Database Developments and Knowledge Discovery Techniques

Although Knowledge discovery in databases are widely accepted technological advancements, their life is just fifteen years [12] [13]. The database development stages during past 40 years are shown in Table 2.1

The gradual development of technologies may be acknowledged in accordance with various time periods described in subsequent sections:

2.2.1 Artificial Intelligence (1950s)

The origins of Data Mining may be traced since 1950s along with the starting phases of Artificial Intelligence (AI) [13][14]. During this period, developments in Pattern Recognition and Rule Based Reasoning were providing the foundation stones for Data Mining [14]. Although during this time the term Data Mining was not established but several techniques were already in use [12].

E Governance Data Center, Data Warehousing and Data Mining: Vision to Realities, 11–30.

Table 2.1 History of Database Developments and Knowledge Discovery Techniques

Stages	Technologies	Product Providers	Characteristics
Artificial Intelligence (1950s)	Pattern Recognition and Rule based Reasoning	IBM	Efficient Methods for Scientific Applications
Data Collection (1960s)	Computer Tape and Disks	IBM, Oracle, Microsoft	Static Data Delivery
Database and RDBMS (1980s)	RDBMS and SQL	Pilot, IRI, Arbor, Redbrick	Dynamic Data Discovery at Record Level
Data Cube, Data Warehouse (1990s)	OLAP, Multi Dimensional Data Analysis	Oracle, IBM, Microsoft	Dynamic Data Discovery at Multiple Level
Knowledge Management (after 1996)	Knowledge Base	Oracle, IBM, Microsoft	Efficient Knowledge Utilization in Large Applications
Knowledge Discovery in Databases (1999)	Large Database based applications	Oracle, IBM, Microsoft	Dynamic Data Discovery and Useful Information
Data Mining (2000)	Advance Algorithm and Massive Data Sets	Oracle, IBM, SAS, Clementine	Discover hidden patterns from huge datasets

2.2.2 Data, Database and RDBMS (1960s–1980s)

Development of storage tapes (1949) and disks (1956) promoted the use of databases at very initial level. In 1960s Relational Database Management System (RDBMS) started to serve commercial organizations for storing large volumes of data. After RDBMS, it was realized that some techniques are required that could be applied and better business profits could be derived [12] [15]. Besides RDBMS various innovative algorithms like Neural Networks, Regression analysis, Likelihood estimates and other types of classification models have been proposed during 1960s [16]. In the beginning of 1970s, different types of DBMS and Information Retrieval (IR) systems were developed.

In 1971, an Information Retrieval method was proposed by Gerard Salton [13]. The proposed smart Information Retrieval system was based on an algebra-based Vector Space Model (VSM). These models would establish main framework of the Data Mining toolkit [16]. Years 1970s, 1980s, and 1990s were the phase of convergence of technologies like AI, IR, Statistics and Database systems along with increased use of high-speed microcomputers, released novel techniques of data analysis and Information Retrieval. At the same time new programming languages were also introduced along with innovative computational methods like Expectation Maximization algorithms, Decision trees and Genetic algorithms [16].

2.2.3 Data Cube and Data Warehouse (1980s–1995)

Initially Structured Query Language (SQL) was used as Information Retrieval techniques but in the middle of 1980s Multi Dimensional Data Analysis and Data Warehouses were introduced for large commercial applications. These applications started to generate large volume of data. In 1990, such Data Warehouses were developed which may store large databases. The Data Warehouse was combination of operational and transactional database in structured format or schema [12]. The developments of Data Warehouses were followed by the evolution of On Line Analytical Processing (OLAP), Association Rule Algorithm, Decision Support Systems (DSS) and Data preprocessing [15] [16]. A Data Warehouse Model is used for multi-dimensional data analysis and represents data as a data cube. Figure 2.1 shows a Multi Dimensional representation of a sales data cube.

A Data Warehouse is a storage space where facts have been stored in well-organized manner for efficient querying, easy access and analysis [14]. A Data Warehouse consists wide range of data items related to any application

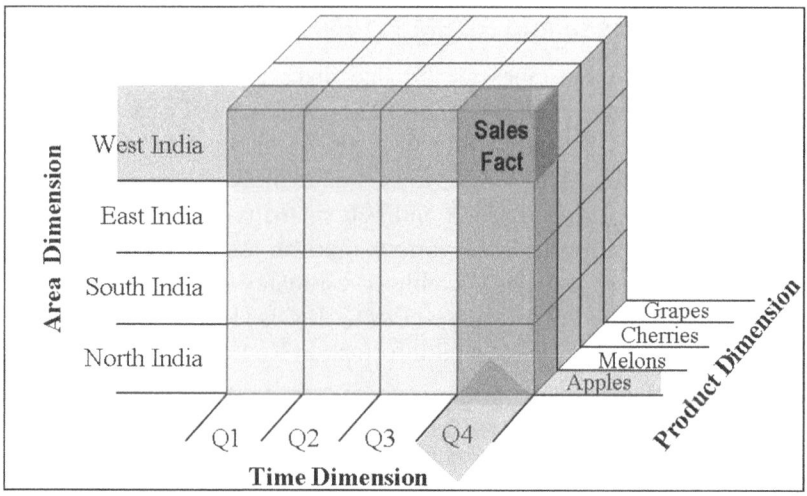

Figure 2.1 A Multi Dimensional Data Cube

[17]. The records in a Data Warehouse could be collected from different distributed, independent and diverse data sources. Any large organization produces day to day operational data which is used in its various business activities.

The operational data are extracted from business databases and converted into a common model. The developed common model has been further added to the Data Warehouse [18]. Figure 2.2 describes structure of a department level Data Marts and a Central Data Warehouse.

The primary gain of using a Data Warehouse is faster query processing and efficient data analysis for better decision making. Data warehouse technologies have been effectively adopted by several industries such as stock, manufacturing, retailing, finance, shipping, communications, healthcare, education and banking etc.

2.3 Features of Data Warehousing

There are following features of Data Warehousing [19].

2.3.1 Subject Orientation

Data Warehousing is subject-oriented, so that it could eliminate data redundancy, which often creates a great problem in application systems [19]. In

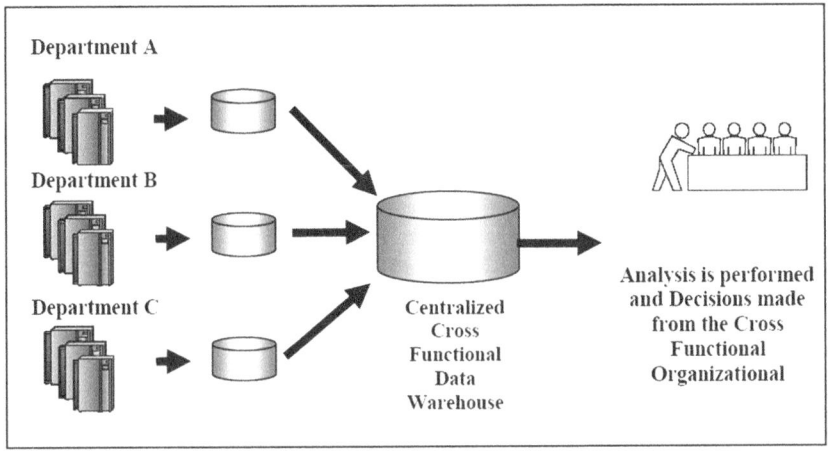

Figure 2.2 Department Level Data Marts and Central Data Warehouse

operational systems, data is arranged according to sequential process of business resulting in data redundancy may occur. For example, "customer" as a record may be presented by different tables in databases, however in Data Warehouse, there may be only one customer object across the whole system.

2.3.2 Integration

Data Integration is an essential characteristic of a Data Warehousing system. It shows error free integrated data within the working system [19]. It may appear as universal naming conventions. It also includes steady measurement of variables, reliable encoding compositions and consistent physical data attributes.

2.3.3 Time Variation

Generally, a Data Warehouse is built for long term data storage and on the other hand the time span involved in operational environment is small [19]. Thus Data Warehouse has in built components of time data as one of its dimension.

2.3.4 Non Volatile

In general all operational system stores data for a small duration, for example on weekly or monthly basis. For such cases data is accessed as well as updated regularly and changing all the time. In contrast to this, in Data

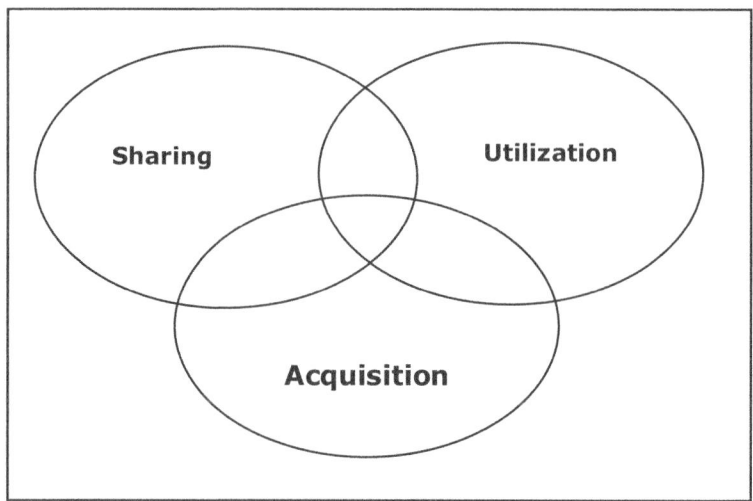

Figure 2.3 Types of Knowledge Management

Warehouse, data is loaded and accessed over the time but historical data remain unchanged [19].

2.4 Knowledge Management (After 1996)

The idea of Knowledge Management (KM) is conceived in the year of 1990s specifically after 1996. The growing IT industries employed Knowledge Management for efficient distribution of valuable information [20] [21]. The growth and formation of insights, expertise and relationships is known as Knowledge Acquisition [22]. Knowledge Sharing is disseminating information and making accessible, which is already known. Distribution of information is significantly enhanced by World Wide Web both within and outside the organizations [23]. Figure 2.3 explains different types of Knowledge Management.

Knowledge Management provides the base to get better working trends, success, and prompt response by sharing process. Knowledge Utilization consists motivating factors to use knowledge and discover better way of implementation. It indicates applications and acceptance of knowledge in various fields of working. Knowledge has been represented in two categories [24]. Explicit Knowledge reflects documented knowledge in the form of books, documents and manuals. Tacit Knowledge refers knowledge that remains in a person's brain, such as wisdom and understanding. It is an individual knowledge and hard to express [24].

2.5   Knowledge Discovery in Databases (1999)

Knowledge Discovery is a process of discovering unaccepted, interesting and valuable data patterns. These data pattern are useful to make predictions or classifications for future use. The data patterns are also helpful to discover innovative business rules and to extend better decision making capabilities of any organization. In advance database application, Decision Support System (DSS) uses traditional query languages, which are difficult, time-consuming and are not quite feasible [25]. An efficient Knowledge Discovery process uses historical data to enhance decision-making capabilities. There are following phases of Knowledge Discovery in Databases as shown in Figure 2.4

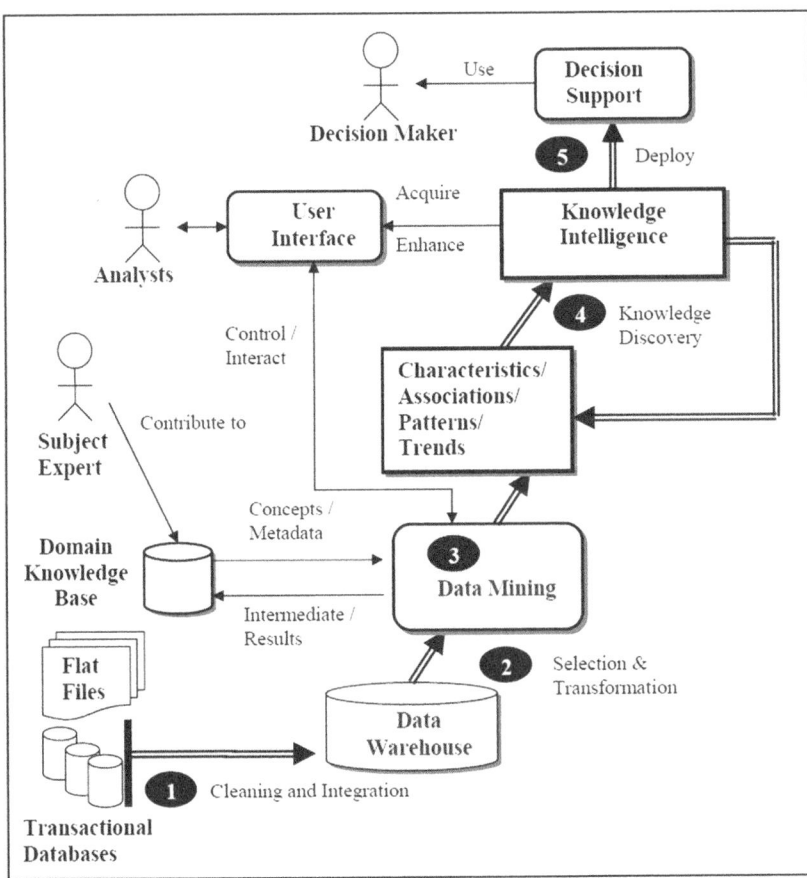

Figure 2.4 Phases of Knowledge Discovery in Databases

- In initial phase, working methodology of any organization is studied with the help of related earlier knowledge. The outcome of this phase is identification of problem, target to be attained and also the causes for success and failure. This also analyzes present working trends in practice, inventory resources, constraints, costs, benefits and losses [26].
- In second phase data is selected from system and target data set has been created according to the identified goals.
- Data cleansing and pre processing are performed in third phase in order to take away noise, outliers, incomplete information and establish data integration and uniformity.
- Fourth phase is used to implement efficient data reduction and projection by using subset selection, attribute construction, aggregations and summarizations [26].
- The final phase includes selection of efficient Data Mining tasks like classifications, segmentations, deviation detections and link analysis [27].

2.6 Data Mining (2000)

During the 1990s, Data Mining techniques became a standard business practice and requirements of most of the commercial organization for effective utilization of their databases. The main cause of this technological revolution is decreasing cost of storage disks and increasing processing power. Both factors proved as catalysts for Data Mining developments. Businesses application was started using Data Mining in Customer Relationship Management (CRM) for obtaining new clients, ever-increasing revenue from existing clients and retaining well clients [28].

Data Mining as an emerging technology works with huge data and covers different techniques of data processing. Data mining methods need huge computations, which are useful to discover interesting patterns existing within large data sets. In recent 10 years, the expansion of computer based business information systems with commercial databases; require proper data understanding and storage for efficient business applications [16].

Traditionally, the primary functions of database systems development are to process operational data generated by business systems. Operational data are collected from several resources and further subjected for processing. The operational data may be of individual account details or any record regarding an existing sales order. The operation data represents the overall details and status of any existing system, which could be further utilized by system analysts [17]. Moreover, every transaction collects and processes the data on back end. This shows that every operation that occurs, for example credit/debit operations of an account, is confined into a record in a database.

Another data processing tool is On Line Transaction Processing (OLTP), which handles with raw data collected from day to day operation. The database either operational or transaction type are very helpful for gathering information of business activities on everyday basis, but does not offer a methodical way to perform historical or trend analysis for large business related Decision Support Systems (DSS) [14] [17]. Predictive modeling is most important approach of Data Mining technology. For example, Data Mining generates predictive models automatically, which forecast how much profit prospects are there in any system and also the risk factor implicated with the system in terms of scam, economic failure, charge-off and related problems.

2.7 Phases of Data Mining

Any Data Mining approach is having fixed set of phases, which are well-ordered and clearly explain various stages of development. Cross Industry Standard Processes of Data Mining (CRISP-DM) is an industry tool, having efficient Data Mining modeling techniques [29]. This model promotes high quality work practices and provides an arrangement to understand improved and quicker outcome of Data Mining. CRISP-DM classifies the Data Mining method into six phases.

Business understanding phase is an initial phase including deep study about the system to finalize the project goal in terms of organizational interest. Data Mining problem definition is finalized in this stage and an initial plan is developed to achieve the objectives.

The data understanding phase primarily includes data collection. The Data expert study about the data to increase acquaintance with the data recognizes deficiencies, determine preliminary insights within data and detect attractive subsets to form hypotheses about unseen information. The Figure 2.5 indicates various phases of Data Mining.

Data preparation phase is the third phase, which provides step by step options to build the final version of dataset. This could be derived from raw data and given to the modeling tool. The different data preparation steps are: data selection, data cleaning, data modeling, integration and formatting.

The fourth stage is modeling in which variety of modeling techniques are chosen, applied and their constraints are adjusted to the best values. Modeling steps contains the choice of the modeling procedure, test design making,

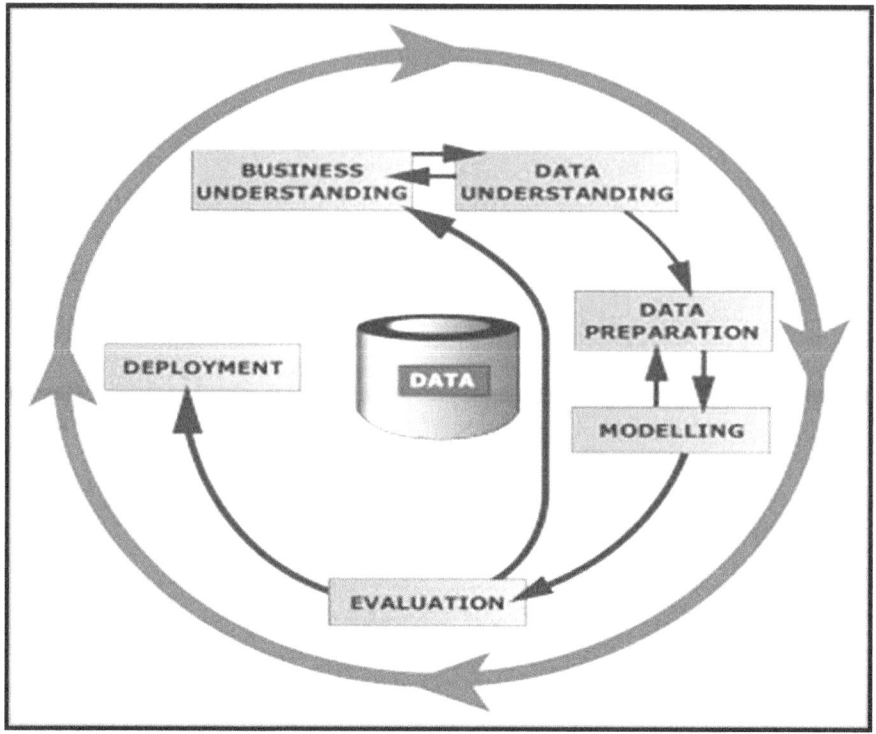

Figure 2.5 Phases of Data Mining

model formation and evaluation [29]. Evaluation is the fifth phase, which systematically evaluates the model and reviews that the models construction achieves the business objectives or not. The last phase is deployment, which ensures certainty that the knowledge achieved must be restructured and worth utilized by the real time systems [29].

2.8 Types of Data

Data Warehouses are utilized by various Data Mining operations with several types of information repositories such as internet and business related databases. Data Mining also involves all types of Databases like text, image, audio, video, temporal, spatial and object oriented database. There may be common or specific Data Mining techniques appropriate for a particular kind of database [27] [30].

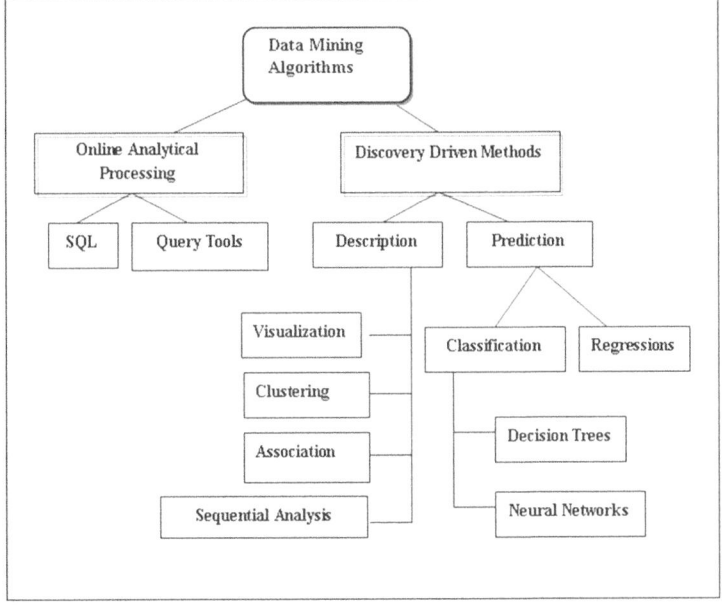

Figure 2.6 Types of Data Mining Algorithms

2.9 Types of Data Mining Algorithms

Important Data Mining algorithms are Classification, Clustering and Regression used to determine future pattern, dependencies and outlier associated with the models. Clustering and Classification are most popular Data Mining algorithms. Data Mining may provide several methods, important steps described below [15]. Different algorithms of Data Mining methods are summarized in Figure 2.6.

2.9.1 Classification

Classification divides data items into target categories or classes. The purpose of classification is correct prediction at the target class for each data points. For example, literacy rate could be predicted as "poor", "average" or "high" with the help of data classification. It is a supervised learning approach with known class categories . Binary and multilevel are the two methods of classification. In binary classification, two possible classes for example, "high" or "low" literacy rates may be considered. Multiclass approach has more than two targets for example, "poor", "average" and "high" literacy rates.

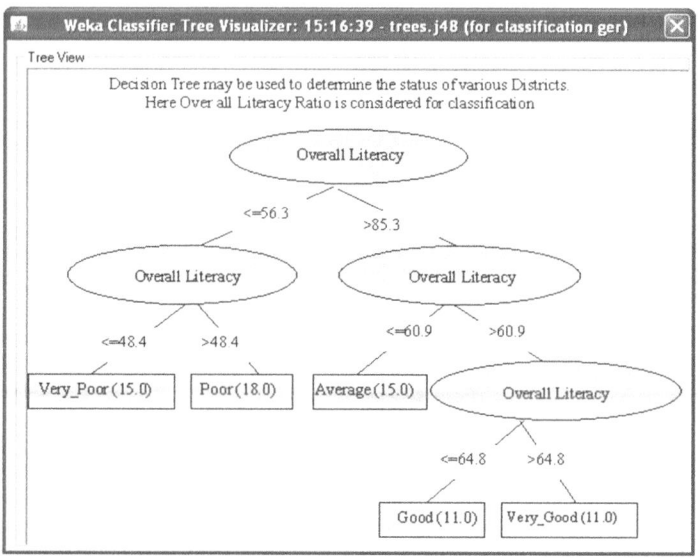

Figure 2.7 Decision Tree

For classification data set is partitioned as training and testing dataset. With the help of training dataset it explores the value of predictors and corresponding value of target. Correctness of the classifier could be tested with the help of test dataset. Some well-known classification algorithms include Decision Tree, Fuzzy Logic, Neural Networks and Support Vector Machines [15] [16] [31].

Decision Trees area type of predictive classifier that uses tree-like graphs as shown in Figure 2.7. With the help of Decision Tree decision makers can choose the most excellent option of several available courses of action [25].

The major benefit of Decision Tree is minimization of ambiguous complicated decisions and it assigns correct values to the results of various actions. Single traversal from root to leaf may indicate a unique class separation based on maximum information gain [32]. The advantages of Decision Tree are simplicity, robustness and quick analysis. The disadvantages of Decision Tree are over fitting; output attribute must be categorical and limited to one output attribute only. Stability is another difficulty with Decision Tree algorithms. Decision tree could also be as one of techniques for multilevel classification [15].

The information processing system of a Neural Network (NN) based Data Classification is motivated by the biological nervous systems having multiple

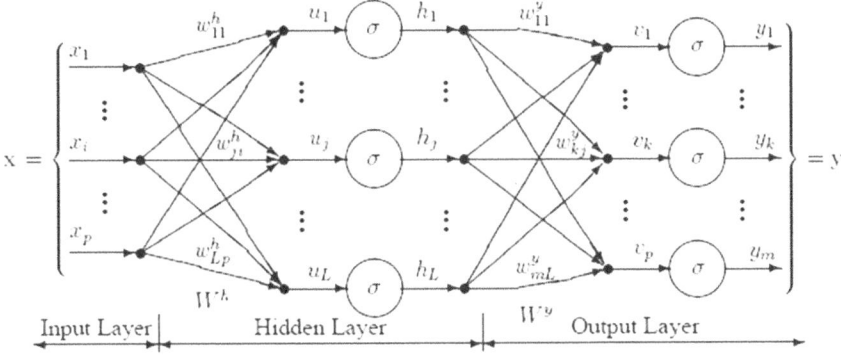

Figure 2.8 Different Layers of Neural Networks

interrelated processing elements known as neurons, functioning in unity in order to solve specific problems [33]. It is adaptive in nature because system changes its structure and adjusts its weight to minimize the error. Regular tuning of weight is based on the information that flows internally and externally through the network at the time of learning phase. Usually Neural Networks are applied to model with complex associations between inputs and outputs and discover pattern in data. Modern Neural Networks are statistical data modeling tools that are used for non-linearly data.

In Neural Networks multiclass problem may be addressed by using multi-layer feed forward technique, in which N neurons have been employed in the output layer rather using one neuron as shown in Figure 2.8 [16].

The advantages of Neural Networks are less formal statistical training requirement, complex nonlinear relationships detection between dependent and independent variable and ability to detect all possible parallel paths between predictor variables. The drawbacks with Neural Networks are local minima and over-fitting. The processing of Artificial Neural Network is hard to interpret as well as requires high processing time if it is large.

Fuzzy logic is closely related to human logic systems proposed by Zadeh in 1965 [34]. In Fuzzy logic interpretation of domain is not precise but it is based on approximation as indicated in Figure 2.9. A fuzzy system is basically a collection of statements based on specific domain real time problems. The domain specification could be shown as "if-then" rules. Currently Fuzzy logic is broadly acknowledged; it could be used in simple, small, embedded micro-controllers and also in large, multi-channel PC, networked and workstation-based control systems. Fuzzy logic is slightly difficult but

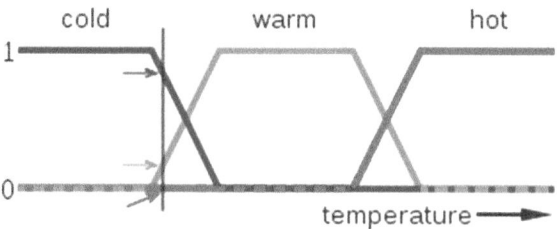

Figure 2.9 Fuzzy Logic

provides definite outcomes even in the existence of indistinct, uncertain, indefinite, noisy datasets [35].

Support Vector Machine (SVM) is a supervised learning approach utilizing Structural Risk Minimization (SRM) induction principle proposed by Vapnik [36]. By using SVM one could easily separate two classes and find out an "optimal" decision surface. SVM could perform classification with linearly separable and non separable data types. Using the Kernel function nonlinearly input data have been mapped into higher dimension feature space so that separation becomes easier and proper training may be done to reduce the structural risk. It offers marginal tolerance on both sides of the decision plane. SVMs were initially developed for binary classification but it could be efficiently extended for multiclass problems [37][38]. While dealing with multiclass SVMs, number of classes and size of datasets are the two prime concerns and going to affect the degree of optimization. Four distinguish techniques are used for multiclass SVMs, i.e., One-against-all, pair wise, error controlled output and all together SVM [39].

2.9.2 Regression

Regression is a statistical method which investigates relationships between variables. By using Regression dependences of one variable upon others may be established. For example price of a product may be estimated according to product demand. This interconnection may be established by applying Regression with datasets [40] [41]. For example, performance analysts of faculty members may be done by Regression. Here, yearly increment and bonus may be estimated as a function of variables such as year of work experience of faculty, qualification of faculty and success rate of courses taught. These variables are known as rating factors because they are used by institutions when settling increments and bonus of the faculty.

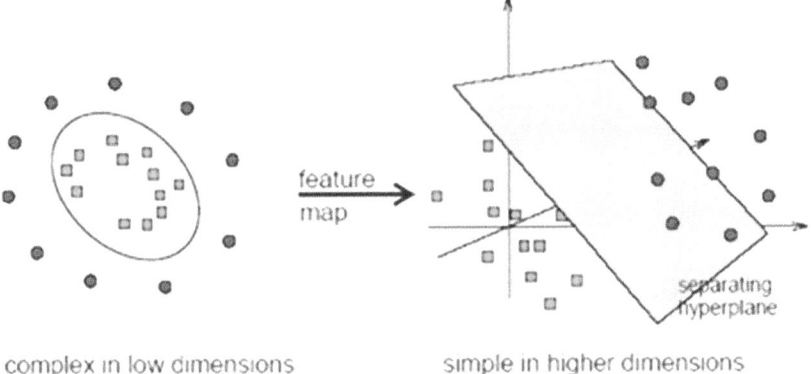

complex in low dimensions   simple in higher dimensions

Figure 2.10 Support Vector Machine

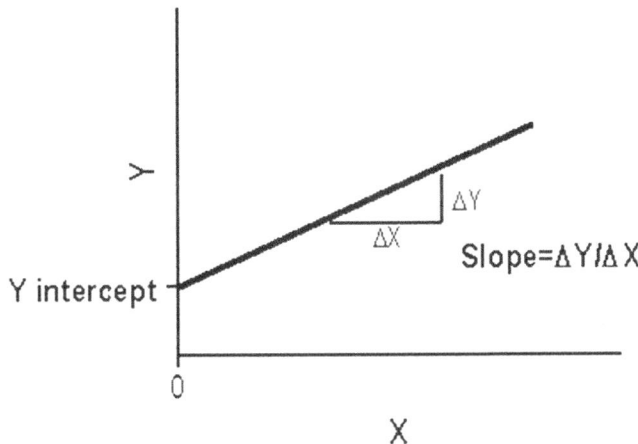

Figure 2.11 Linear Regression

Linear Regression is a statistical method which correlates a dependent variable with independent variables. The model is constructed using linear functions. In this approach dependent and independent variables are already known and the purpose is to spot a line that correlates between these variables [42]. Figure 2.11 shows an example of Linear Regression.

A Linear Regression function is represented as Yp= mX + b, here, Yp is termed as dependent variable and X termed as independent variable , m is slope and b is constant value which determines Y-intercept. Regression is limited to numeric data only and cannot work with nominal dataset such as excellent, good, average and poor [42].

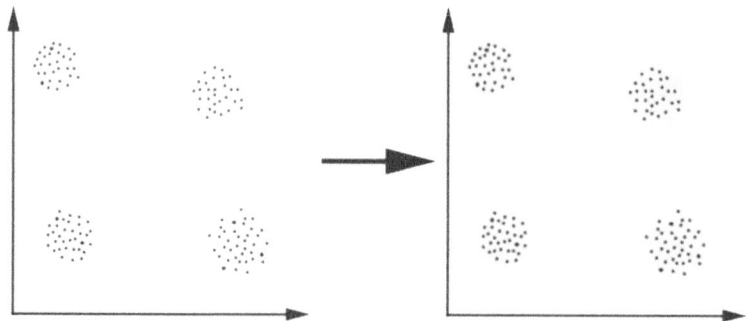

Figure 2.12 Data Clustering

2.9.3 Clustering

In Clustering large databases are separated into the form of small different subgroups or clusters. The clusters could be mutually exclusive, overlapping or hierarchical [43]. Clustering is unlike to Classification since it has no predefined classes. Clustering approach of Data Mining is useful to identify similarity in the system. Each data point within a cluster has similar properties and there must be large dissimilarity to other clusters. Clustering is a way of data partitioning. In K-means clustering objects may grouped into a tree of clusters. Grid based model and density-based method models are also used for clustering [43].

Outlier detection is also a way of clustering in which those items have been considered which do not fit in any cluster [43]. Sometimes these objects represent errors in the data, sometimes represents the most interesting patterns. Market Basket Analysis is used to learn the things which may go together [44]. It is interesting because it provides unanticipated associations, for example, Market Basket Analysis of non-resident student's records might reveal that, besides admission to the college, the students may take admission in hostels and avail mess facilities. The institution could use this information in arranging fooding and lodging according to similar background students. Figure 2.12 indicates data clusters.

In above case four data groups could be easily identified and distance parameter is used to establish similarity. Data objects belong to same cluster having same distance measure. This is known as Distance-Based Clustering [45]. There are many problems with Clustering. Some are:

- A single Clustering algorithm does not address all the requirements adequately.

- Dealing with high dimensions and huge data items become challenging because of time complexity.
- A clear distance measure should be defined which is difficult with high dimensional space.
- Result Interpretation of clustering is crucial and possible in several ways.

Variety of clustering techniques exists and choice of appropriate clustering approach depends on the nature of data. In K-Means Clustering, input parameter has been taken and it partitions the dataset in different clusters. In Clustering similar data points assigned to one cluster and theses points must be dissimilar to other cluster points [46] [47]. Each cluster must have a cluster centre. For example a cluster center k can represent the clustering of M points and number of cluster centre should be less than M data points. Clustering initiates with M random points and each of these M points belong to any cluster K. Initially selection of cluster centre k is random. For every iteration, a data point is handed over to the cluster based on similarity of cluster mean the distance between the data points [46] [47]. The latest mean is calculated and this step is recurred to accommodate every newly arrived data points. The approach is intended to form compact clusters of similar data points with fare dissimilarity with other clusters. Cluster similarity could be characterized in the form of cluster mean which is also considered as centroid of the cluster.

K means is a simple, efficient, scalable and less complex approach. It is a self organized approach and easily initiates clustering process so many complex clustering approach uses K means as beginning process. The limitation of K means approach is lack of information regarding choice of initial parameter and its inability to handle non linear pattern of datasets [46] [47].

Density based algorithms identify clusters as per its density levels. A density-based grid unit has been used and a high dimensional feature space is divided into several sections by using density-based grid units. Cluster centers could be located at high density areas and cluster boundaries may be of low density areas. So high density regions could be estimated as ideal clusters and low density area may also represent outliers or noise data points.

2.10 Status of Emergence of E Governance

E Governance services in India have been enormously admired since last few years because it has very large target segments of citizens, governments and business parties. E Governance operates at the intersection among ICTs and Government processes. E Governance could be alienated into three

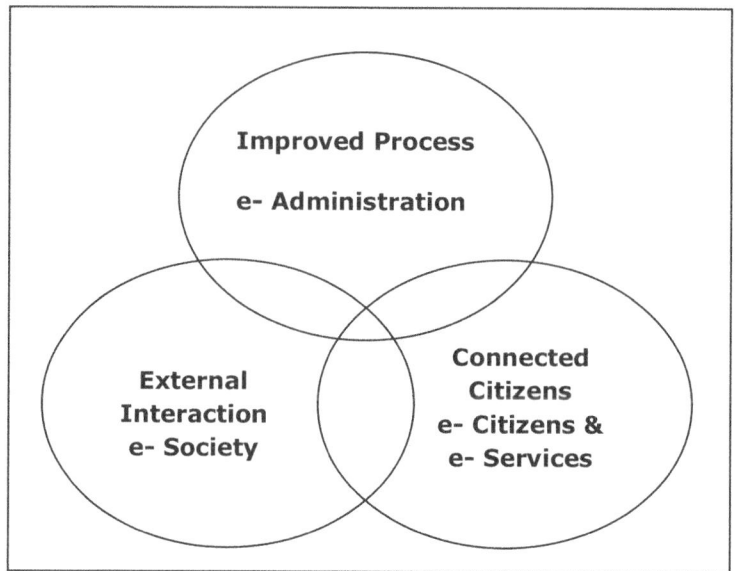

Figure 2.13 Domains of E Governance

overlapping spheres: E-administration, E-services and E-society. The Figure 2.13 indicates three domains of E Governance [48].

E administration refers to several mechanisms which convert paper processes of traditional offices into electronic processes. The E Administration is helpful to achieve full accountability and transparency leading to an improved E Governance. E Services includes electronic media or Internet based services. An E Society is a cluster of individuals having common interests, distinguished traditions and association by means of internet [49].

2.10.1 Definition of E Governance

In straightforward way, the E Governance may be characterize as giving citizens an alternative of Anytime, Anyhow, Anywhere (AAA) access to government services, consequently influencing the State, Central Government, to reach a most important position in Information Technology. Gartner Group defines E Governance as an attempt towards constant optimization of government service deliverance, citizen participation, proficient technologies, internet, and new media [50]. According to Mark Forman, E Governance is making use of WWW standards for efficient and well-organized service delivery [51].

2.10.2 Phases of E Governance

Different Governments have diverse strategies to implement E Governance initiatives. According to Gartner, the phases of E Governance are categories as follows: The first stage of advancement is emerging phase and it is the concerned with delivery of information [52]. The interactive phase offers interactive web based programs with improved features, to simplify and automate government functions. The transaction phase is responsible to develop an efficient interface so that citizen may complete any task online at any time. Theses task are like online enquires, license renewals, paying taxes, downloading forms, submitting fees and other utility bills. In the seamless phase fully integrated E services across administrative boundaries should be established. The most important phase of E Governance is transformation in which overall renovation of the government activities could be performed [53].

2.10.3 Scope of E Governance

Even though E Governance covers an extensive range of action plans and stakeholders, three different segments could be identified [2.38]. Government-to-Government (G2G) E Governance includes data sharing and performs electronic exchanges among governmental departments. These involve both intra and inter department information sharing at all levels. Government-to-Business (G2B) is an online interaction between government and business agencies, for reducing costs throughout better procurement practices, improved competition and makes more efficient regulatory processes. The Government-to-Citizen (G2C) is planned to make simple citizen government interaction, which is the prime purpose of E Government [52] [53].

2.10.4 Current Status of E Governance

Currently, the situation of E Governance in India represents a broad dissimilarity in computerization and ICT usages within and outside the government. Although there are continuous efforts in the expansion of E Governance services, the entire government activities, especially in the states having poor ICT infrastructure.

In India internet access is improved in previous years and showing rising pattern [54]. The increased internet users indicate best media reachability in masses with least effort. But still the growth is not satisfactory in case of teledensity. India still has poor penetration of telephone lines which is the first requirement of E Governance infrastructure [54].

2.11 Conclusion

This chapter indicates that the various Information Technology trends like Data, Database and Database Management System (DBMS), created a scenario of "Information Rich and Knowledge Poor". Clustering, Classification and Regression are the three major techniques to perform Data Mining operations with real datasets. Support Vector Machines (SVM) is an efficient classification approach based on maximum margin hyper-plane. It shows less classification error and an efficient model. Originally SVM performs only binary classification. But a combination of Kernel function has been used to create high dimension feature space for data points so that non linear classification could be achieved by using SVM.

There are several E Governance projects ongoing in conjunction with Data Mining applications. Majority of the projects are aimed to cover only a specific area like Health, Education or Finance. The E Governance application domain is not confined to a individual application but it covers broad area of spectrum under a single umbrella. So there is an imperative requirement of a complete scheme for Data Mining and Data Warehousing based E Governance model, which could cover E Governance activities in totality.

3

World Wide Status of E Governance

3.1 Introduction

In present scenario E Governance initiatives have been considered beneficial in terms of quicker citizen service, better efficiency, accountability, transparency and high speed connectivity. These encouraging results of E Governance system motivated not only the developed countries but also the developing as well as least developing countries for adaption of E Governance System. Over a period of time, as the enthusiasm towards E Governance has been increased among all nations, the United Nation started a survey to identify the world wide status of E Governance. The survey was conducted among 192 countries for analyzing their preparedness toward E Governance on the basis of E Governance Development Index. The E Governance Development Index [55] has been proposed by United Nations. It is measured on the basis of preparedness and motivation of a government for adapting usage of Information and Communication technologies to perform citizen services. It is based on following three important parameters as indicated in Table 3.1:

The above parameters are important to consider because availability, accessibility and citizen participation are very significant in E Governance. The maximum value of all the above parameters has been taken as 1 and minimum value is considered as 0. Following formula is used to calculate E Governance Development Index [55]:

3.1.1 Global Status of E Governance

As the development of E Governance is a major concern all over the world, the E Governance Development Index (EGDI) could be chosen to rank various cities globally. Figure 3.1 shows ranking of most populous countries on the basis of E Governance Development Index. The Figure indicates that China and India are the two most populous countries with improving pattern on E

E Governance Data Center, Data Warehousing and Data Mining: Vision to Realities, 31–72.

Table 3.1 E Governance Development Index

Three Dimensions of E Governance Development Index (EGDI)	Features
Online Service Index	It Ensures online availability of government information through websites. It could be managed and accessed by Web Content Accessibility Guidelines.
Tele Communication Infrastructure Index	It includes five different modes of tele-connectivity and internet accessibility. • number of personal computers per 100 persons • number of Internet users per 100 persons • number of telephone lines per 100 persons • number of mobile cellular subscriptions per 100 persons • Number of fixed broadband subscribers per 100 persons.
Human Capital Index	The human capital index is a based on two parameters: Adult literacy rate the combined primary, secondary, and tertiary gross enrollment ratio.

Governance Development Index [56]. It is also clearly established that United States and Japan are world top rankers on the basis of their EGDI.

$$EGDI = (0.34 \times \text{Online Service Index}) + (0.33 \times \text{Telecommunication Infrastracture Index}) + (0.33 \times \text{Human Capital Index})$$

The 20 best economies of the world are shown in Figure 3.2, on the basis of the development of E Governance. The range of the E Governance Development Index for these 20 countries ranges from 160% to 190% of the average of the index of whole world. The world leader republic of Korea maintains 0.9283 E Governance Development Index and represents East Asia in the world. Europe is highlighted in 2nd, 3rd, 4th positions and represents top most rankings [57]. USA is at the 5th position and is showing sustainable growth. Australia and New Zealand are oceanic countries showing their presence among the top 20 countries. The development of E Governance is prevailing in western Asia which is marked by Israel. The following Figure 3.2 shows the ranking of top 20 countries and next 20 countries following top 20 countries.

In Figure 3.2, apart from top 20 countries, the next 25 countries following them are mainly from Europe, which are 16 in number. It is also clear from the Figure that 6 Asian countries and 3 American countries are

Country	E-gov. development index		World e-gov. development ranking		Population (in millions)
	2012	2010	2012	2010	
China	0.5359	0.4700	78	72	1,341
India	0.3829	0.3567	125	119	1,225
United States	0.8687	0.8510	5	2	310
Indonesia	0.4949	0.4026	97	109	240
Brazil	0.6167	0.5006	59	61	195
Pakistan	0.2823	0.2755	156	146	174
Nigeria	0.2676	0.2687	162	150	158
Bangladesh	0.2991	0.3028	150	134	149
Russian Federation	0.7345	0.5136	27	59	143
Japan	0.8019	0.7152	18	17	127
Mexico	0.6240	0.5150	55	56	113

Figure 3.1 EGDI in Largest Population Countries
(SOURCE: United Nation E Governance Development Database 2012)

in the next 25 counties. In this group of 25 countries Austria (0.7840) is in top position and Cyprus (0.6508) is in last position. For the deployment of information technologies there is a need of good infrastructure that facilitates the utilization of E Governance services. Lack of knowledge, poor infrastructure and human capital limitation are the main problems for the implementation of E Governance services. The whole world is divided into five different segments for analyzing the growth of E Governance services as indicated in Figure 3.3. Europe and America have good E Governance Development Index as compared to other regions and the condition of Africa in the development of E Governance service is very poor as indicated in Figure 3.3. It is also noted that most of the population of world are living in Asia but E Governance Development Index is not as good as other regions.

Figure 3.4 shows the positions of EGDI for different continents from the year 2003 to 2012. The figure represents the growth of E Governance practices across five regions of world such as Europe, Asia, Africa, Oceanic and America and among all these regions Europe has secured top position in the world due to their enhanced E Governance services as compared to others.

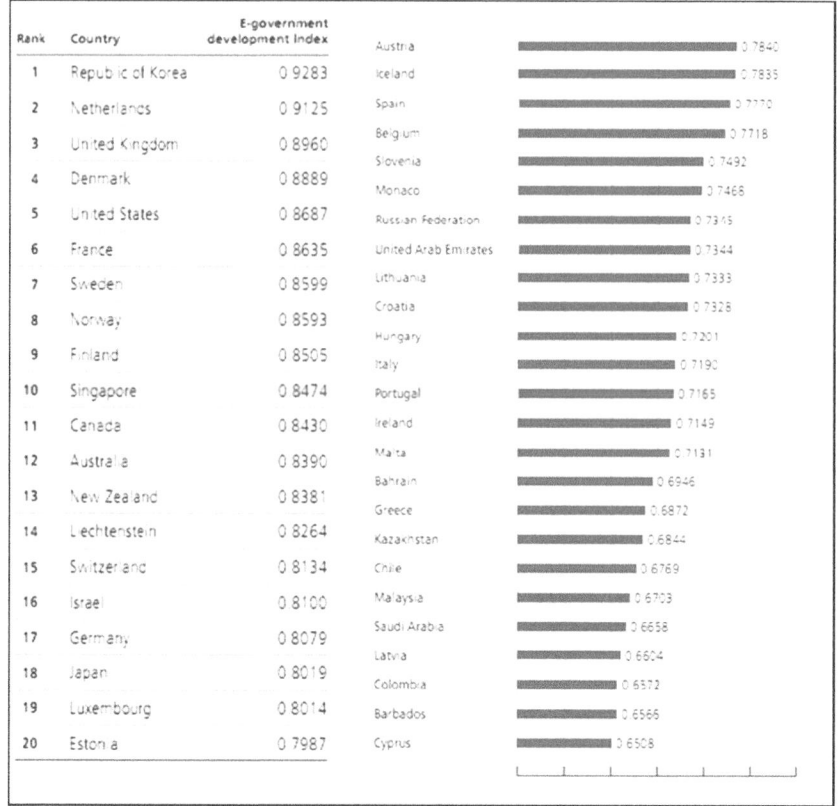

Rank	Country	E-government development Index
1	Republic of Korea	0.9283
2	Netherlands	0.9125
3	United Kingdom	0.8960
4	Denmark	0.8889
5	United States	0.8687
6	France	0.8635
7	Sweden	0.8599
8	Norway	0.8593
9	Finland	0.8505
10	Singapore	0.8474
11	Canada	0.8430
12	Australia	0.8390
13	New Zealand	0.8381
14	Liechtenstein	0.8264
15	Switzerland	0.8134
16	Israel	0.8100
17	Germany	0.8079
18	Japan	0.8019
19	Luxembourg	0.8014
20	Estonia	0.7987

Country	Value
Austria	0.7840
Iceland	0.7835
Spain	0.7770
Belgium	0.7718
Slovenia	0.7492
Monaco	0.7468
Russian Federation	0.7345
United Arab Emirates	0.7344
Lithuania	0.7333
Croatia	0.7328
Hungary	0.7201
Italy	0.7190
Portugal	0.7165
Ireland	0.7149
Malta	0.7131
Bahrain	0.6946
Greece	0.6872
Kazakhstan	0.6844
Chile	0.6769
Malaysia	0.6703
Saudi Arabia	0.6658
Latvia	0.6604
Colombia	0.6572
Barbados	0.6566
Cyprus	0.6508

Figure 3.2 Top 40 Countries as per their EGDI

(SOURCE: United Nation E Governance Development Database 2012)

After Europe region, America is second best and achieved better performance in the development of E Governance services by offering online services to its citizen. It is also noted that the growth rate of Asia is average. The development rate of E Governance services in Africa is very poor among all the region because of poor availability of Information and Communication Technology enabled infrastructure.

3.2 Status of E Governance in Africa

The E Governance development status of Africa is observed in five different geographical segments as shown in Figure 3.5.

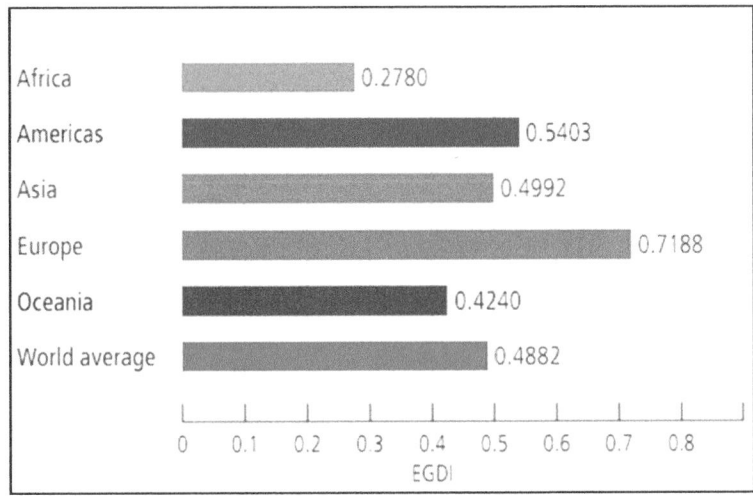

Figure 3.3 EGDI According to Geographical Area
(SOURCE: United Nation E Governance Development Database 2012)

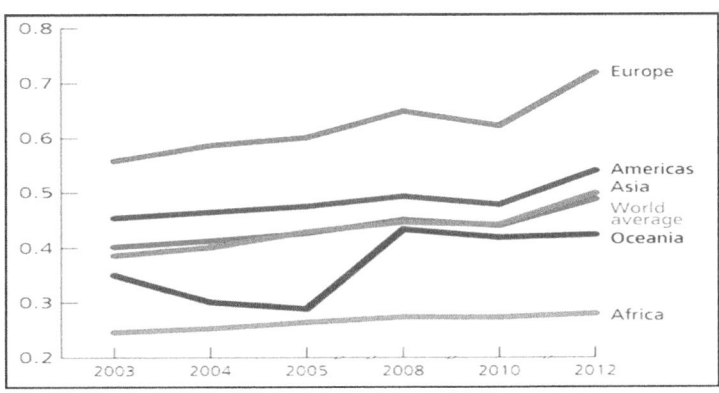

Figure 3.4 Position of EGDI for Different Continents from 2003 to 2012
(SOURCE: United Nation E Governance Development Database 2012)

Comparative study of E Governance practices of various countries in Africa for the year 2010 and 2012 are mentioned in Figure 3.6. Seychelles is improving its rank in the global scenario and also securing 1st rank among the African countries. Most of the other countries in Africa region such as South Africa, Tunisia and Egypt have declined their rank in the global scenario for 2012 [58].

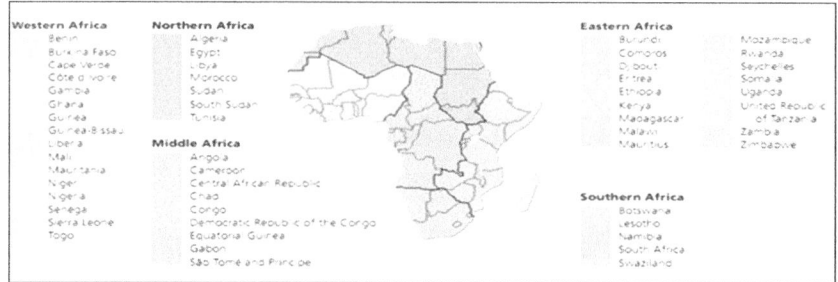

Figure 3.5 Sub Regions of Africa

(SOURCE: United Nation E Governance Development Database 2012)

The challenges for the development of E Governance services in African Countries are due to lack of infrastructure and poor knowledge. There is no proper implementation of digitization services in most of the countries. It is also remarkable that infrastructure is not sufficient enough to provide online services to its citizens. Due to all these factors Africa is not experiencing the development of E Governance facilities and it is least developed as compared to other continents.

E Governance growth status of Africa from the year 2008 to 2012 is shown in Figure 3.7. As figure indicates that Southern Africa has achieved greater performance in the development of E Governance practices while the performance of Western Africa is poor among all sub regions of Africa. Country wise details of E Governance development status in Africa is described in subsequent sections.

3.2.1 Eastern Africa

Seychelles (0.5192) has secures top rank in the sub region of Eastern Africa and also improves its rank in the global scenario as shown in Figure 3.8. Here the development of E Governance is better because the country has better telecommunication services, awareness towards health care, fair economic condition and quality education.

Mauritius maintains 2nd rank in the sub region of Eastern Africa with 0.5066 EGDI by offering various E Governance services such as transportation, online tax payment, scholarships and work permit using electronic media. But overall ranking in the global scenario of Mauritius is not good and it slipped

Rank	Country	E-gov. development index		World e-gov. development ranking	
		2012	2010	2012	2010
1	Seychelles	0.5192	0.4179	84	104
2	Mauritius	0.5066	0.4645	93	77
3	South Africa	0.4869	0.4306	101	97
4	Tunisia	0.4833	0.4826	103	66
5	Egypt	0.4611	0.4518	107	86
6	Cape Verde	0.4297	0.4054	118	108
7	Kenya	0.4212	0.3338	119	124
8	Morocco	0.4209	0.3287	120	126
9	Botswana	0.4186	0.3637	121	117
10	Namibia	0.3937	0.3314	123	125
	Regional Average	0.2780	0.2733		
	World Average	0.4882	0.4406		

Figure 3.6 EGDI Trends in African Countries

(SOURCE: United Nation E Governance Development Database 2012)

down in its world ranking as highlighted in Figure 3.8 [59]. It is also observed that instead of maintaining or improving its rank in the global scenario, the positions of African Countries have been decreased.

The following are the recognized E Governance projects of various developing countries of Eastern Africa:

3.2.2 Uganda

- For the development of E Governance policies Uganda government has established an E Government portal.
- To support Uganda parliament there is a project for Parliamentary Technical Assistance using Information and Communication Technologies [60].

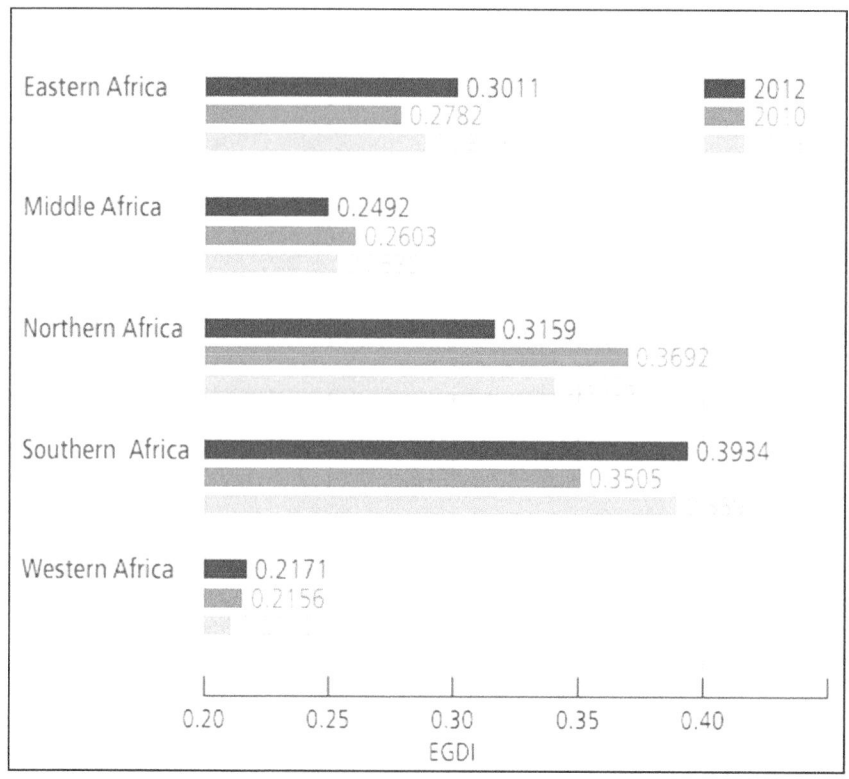

Figure 3.7 E Governance Development Status of Africa 2008–2012

(SOURCE: United Nation E Governance Development Database 2012)

3.2.3 Zambia

- Zambia.co.zm is the E Government portal for Zambia Government that offers various services online [61].
- ZAMLII: This portal collects the documents regarding Zambian government issues and provides secure and easy access to the people [62].

3.2.4 Mozambique

- mozambique.mz is an E Government portal for Mozambique government based operations [63].
- E-SISTAFE: This system is basically monitoring the investment and salaries of public [64].

Country	E-gov. development index		World e-gov. development ranking	
	2012	2010	2012	2010
Seychelles	0.5192	0.4179	84	104
Mauritius	0.5066	0.4645	93	77
Kenya	0.4212	0.3338	119	124
Zimbabwe	0.3583	0.3230	133	129
United Rep. of Tanzania	0.3311	0.2926	139	137
Rwanda	0.3291	0.2749	140	148
Uganda	0.3185	0.2812	143	142
Madagascar	0.3054	0.2890	148	139
Zambia	0.2910	0.2810	154	143
Mozambique	0.2786	0.2288	158	161
Malawi	0.2740	0.2357	159	159
Comoros	0.2358	0.2327	171	160
Ethiopia	0.2306	0.2033	172	172
Burundi	0.2288	0.2014	173	174
Djibouti	0.2228	0.2059	176	170
Eritrea	0.2043	0.1859	180	175
Somalia	0.0640	0.0000	190	N/A
Sub Regional Average	**0.3011**	**0.2782**		
World Average	**0.4882**	**0.4406**		

Figure 3.8 Trends of E Governance Development in Eastern Africa
(SOURCE: United Nation E Governance Development Database 2012)

3.2.5 Ethiopia

- ethopia.gov.et is an E Government portal for Ethiopia government [65].
- Devinet: This is a web based project that provides information and networking gateway for establishing NGO policies [66].
- Ethiopia Government also using a distance learning video conferencing centers for its citizens.

3.2.6 Kenya

- kenya.go.ke is E Government internet based portal for Kenyan government that provide online services to its people [67].

3.2.7 Zimbabwe

- gta.gov.zw is an official Portal for E Government [68].

3.2.8 Malawi

- malawai.gov.mw is an E Government official portal that comprise online services [69].

3.2.9 Djibouti

- Internet based E Government portal republique-djibouti.com for Djibouti Government that provides online services for the development of E Governance [70].

3.2.10 Tanzania

- tanzania.go.tz is an E government portal for enhancing the development of E Governance services for Tanzanian government [71].
- It also has a Government Payroll and Human Resources System (GPHRS) for monitoring the government employee effectively [72].

3.2.11 Middle Africa

Although the position of Middle Africa in the world ranking is not good but still it has improved its ranking during 2010 to 2012 as presented in Figure 3.9 . Among all sub regional countries the rank of Gabon country is highest with 0.3687 EGDI, Sao Tome is in 2nd position and Angola is in 3rd position with EGDI 0.3327 and 0.3203 respectively.

Facebook and Twitter played an important role in the improvement of Gabon Country by involving its citizen [73]. Here people are more concerned towards different issues regarding social welfare such as health, education etc. E Governance Development trends in other African countries are also indicated in Figure 3.9. Following are the established projects in the E Governance of various developing countries of middle Africa:

	E-gov. development index		World e-gov. development ranking	
Country	2012	2010	2012	2010
Gabon	0.3687	0.3420	129	123
Sao Tome and Principe	0.3327	0.3258	138	128
Angola	0.3203	0.3110	142	132
Cameroon	0.3070	0.2722	147	149
Equatorial Guinea	0.2955	0.2902	151	138
Congo	0.2809	0.3019	157	135
Democratic Republic of the Congo	0.2280	0.2357	174	158
Chad	0.1092	0.1235	189	182
Central African Republic	N/A	0.1399	N/A	181
Sub Regional Average	0.2492	0.2603		
World Average	0.4882	0.4406		

Figure 3.9 E Governance Development Trends in Middle Africa

(SOURCE: United Nation E Governance Development Database 2012)

3.2.12 Cameroon

- spm.gov.cm is an E Government portal for citizen services [74].
- E-tax portal for providing guidance about data relevant to tax, establishing various policies for its citizens, providing information regarding payment immediately.

3.2.13 Northern Africa

Figure 3.10 shows the positions of the Northern Africa sub region countries in global scenario based on E Governance Development Index. Tunisia is at top with highest EGDI among different countries of Northern Africa. Algeria is also performed well with 0.3608 EGDI [75] but other countries are not able to maintain their rank globally due to lack of telecommunication infrastructure.

Country	E-gov. development index		World e-gov. development ranking	
	2012	2010	2012	2010
Tunisia	0.4833	0.4826	103	66
Egypt	0.4611	0.4518	107	86
Morocco	0.4209	0.3287	120	126
Algeria	0.3608	0.3181	132	131
Sudan	0.2610	0.2542	165	154
South Sudan	0.2239	N/A	175	N/A
Libya	N/A	0.3799	N/A	114
Sub Regional Average	0.3159	0.3692		
World Average	0.4882	0.4406		

Figure 3.10 E Government Development in Northern Africa
(SOURCE: United Nation E Governance Development Database 2012)

3.2.14 Libya

- Libya has initiated various E Governance steps and the current status of E Governance in Libya is as explained in Table 3.2 [76]:

3.2.15 Tunisia

- E Government portal for Tunisian Government provides easier and faster access to various services based on internet such as home loans, issuance of driving license to its citizens [77] [78].

3.2.16 Egypt

- E Government portal of government provides unique id to its citizens, on-line reservation of bus, online registration of Birth certificate, extraction of online marriage certificate etc [79].

Table 3.2 Status of E Governance in Libya

Presence	Characteristics	Libyan Scenario
Web Presence	Information: Presence	Web sites of Libyan government, departments & Universities available.
Interactive Presence	Interaction Intake processes	Online processes, form filling, tenders, tax deposition are rare.
Transactional Presence	Transaction: Complete transactions	Completion of online processes is totally absent
Networked Presence and F-Participation	Transformation: Integracion& change	Integrated platform for government services and virtual agencies for this arc not available

3.2.17 Sudan

- TOKTEN is an E Governance portal of Sudan that delivers various government services to the public and provides a transparent view about the functioning of government [80] [81] [82] [83].

3.2.18 Southern Africa

Southern Africa has improved its global ranking of E Governance Development Index through online portals. The Figure 3.11 shows that Botswana, Namibia and Lesotho are also maintaining satisfactory EGDI irrespective of their regional circumstances [84] while other countries are not even able to maintain their universal ranking related to E Governance development.

The following are the few established E Governance projects in Southern African countries:

3.2.19 South Africa

- gov.za is an E Government portal provides online access to information regarding various government policies [85].
- Electoral Commission system for voter that provides online registration of voter and verifies their votes [86].

3.2.20 Lesotho

- lesotho.gov.ls is a portal for providing various E Governance services [87].

Country	E-gov. development index		World e-gov. development ranking	
	2012	2010	2012	2010
South Africa	0.4869	0.4306	101	97
Botswana	0.4186	0.3637	121	117
Namibia	0.3937	0.3314	123	125
Lesotho	0.3501	0.3512	136	121
Swaziland	0.3179	0.2757	144	145
Sub Regional Average	**0.3934**	**0.3505**		
World Average	**0.4882**	**0.4406**		

Figure 3.11 Status of E-Government Development in Southern Africa
(SOURCE: United Nation E Governance Development Database 2012)

3.2.21 Swaziland

- E Government portal gov.sz provides secure and easy access of E Governance services to its citizen [88].

3.2.22 Namibia

- grnnet.gov.na is an E Government portal for Namibian government collects the documents regarding Namibia judiciary issues [89] [90].

3.2.23 Botswana

- Botswana E Government portal provides E Governance services regarding health issues, Land management, policies for education, transport services etc. [91].

3.2.24 Western Africa

As indicated in Figure 3.12, Cape Verde is a leading country with development index 0.4297 and progressing well [92]. But most of the countries of Western

Country	E-gov. development index		World e-gov. development ranking	
	2012	2010	2012	2010
Cape Verde	0.4297	0.4054	118	108
Ghana	0.3159	0.2754	145	147
Gambia	0.2688	0.2117	161	167
Nigeria	0.2676	0.2687	162	150
Senegal	0.2673	0.2241	163	163
Côte d'Ivoire	0.2580	0.2805	166	144
Liberia	0.2407	0.2133	169	166
Togo	0.2143	0.2150	178	165
Benin	0.2064	0.2017	179	173
Mauritania	0.1996	0.2359	181	157
Guinea-Bissau	0.1945	0.1561	182	179
Mali	0.1857	0.1815	183	176
Burkina Faso	0.1578	0.1587	185	178
Sierra Leone	0.1557	0.1697	186	177
Niger	0.1119	0.1098	188	183
Guinea	N/A	0.1426	N/A	180
Sub Regional Average	**0.2171**	**0.2156**		
World Average	**0.4882**	**0.4406**		

Figure 3.12 E Government Development in Western Africa

(SOURCE: United Nation E Governance Development Database 2012)

Africa are not doing well because of poor infrastructure. Various established E Governance project of Western African countries are:

3.2.25 Ghana

- E Governance centers are established to help the people by providing information regarding the various policies and objectives of the government [93].

- Environmental Information Network has been developed to provide effective networking among different government agencies.
- African Elections Project provides support for covering elections across the whole countries and also provides online information regarding election [94].
- Community information centers supports internet connection across countries, develop awareness regarding schooling in rural areas etc [95].

3.2.26 Nigeria

- Police diary project allows the citizen to communicate with the police of Nigeria and file their report online [96].
- Information SMS request project allows public to discuss about various political issues with the Nigerian government.
- Touch screen kiosk project for HIV/ AIDS provides awareness about HIV/AIDS to public with the help of interactive screens [97].
- WANGONet project is dedicated to meet the needs of society [98].

3.2.27 Senegal

- Forum civil Website provides information to the public about economy and creates awareness regarding corruption [99].
- Administrative procedures website promotes the digitization of various government services.

3.2.28 Cape Verde

- The E Governance portal of Cape Verde provide support to various Government program and enhance the online delivery of administrative services to the public [100].

3.3 Status of E Governance in America

America is divided into four sub region as shown in Figure 3.13:

The E Governance development growth of Northern America (USA and Canada) is highest with 0.8559 EGDI that is far ahead than the world average EGDI as indicated in Figure 3.14. Entire America with South Africa and Caribbean enhanced its E Governance status in the year of 2012.

There are various E Governance projects running successfully in America, country wise details of few projects are described as follows:

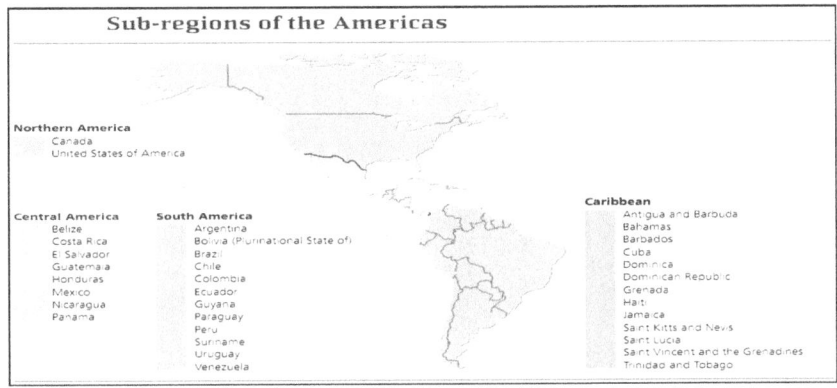

Figure 3.13 Sub-Regions of the America

(SOURCE: United Nation E Governance Development Database 2012)

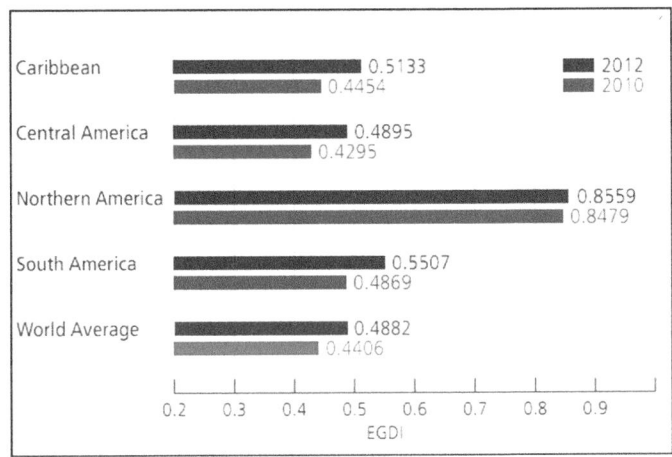

Figure 3.14 Regional E Government in America

(SOURCE: United Nation E Governance Development Database 2012)

3.3.1 Caribbean

According to Figure 3.15, Barbados with 0.6566 EGDI is at the top of among all countries in Caribbean. Antigua and Barbuda (0.6345) and Bahamas (0.5793) have 2nd and 3rd rank respectively. Barbados has user information management system which manages communication between government, citizens and business parties and also facilitates online taxation service [101]. Other

Country	E-gov. development index		World e-gov. development ranking	
	2012	2010	2012	2010
Barbados	0.6566	0.5714	44	40
Antigua and Barbuda	0.6345	0.5154	49	55
Bahamas	0.5793	0.4871	65	65
Trinidad and Tobago	0.5731	0.4806	67	67
Dominica	0.5561	0.4149	73	105
Grenada	0.5479	0.4277	75	99
Saint Kitts and Nevis	0.5272	0.4691	81	75
Saint Vincent and the Grenadines	0.5177	0.4355	85	94
Dominican Republic	0.5130	0.4557	89	84
Saint Lucia	0.5122	0.4471	90	88
Jamaica	0.4552	0.4467	108	89
Cuba	0.4488	0.4321	110	96
Haiti	0.1512	0.2074	187	169
Sub Regional Average	**0.5133**	**0.4454**		
World Average	**0.4882**	**0.4406**		

Figure 3.15 E Government Development in Caribbean

(SOURCE: United Nation E Governance Development Database 2012)

countries are also very keen to improve their E Governance infrastructure especially in telecommunication and human capitalization. In this region digital library system, Info-med health system, environmental management are the key focus areas of E Governance. Established E Governance projects in some developing countries in Southern America are mentioned below:

3.3.2 Bahamas

- Tax portal for making payment of taxes online without going to tax office [102].
- E Government portal provides various services such as online registration of Birth certificate, death certificate, online extraction of marriage certificate and it also provides unique id, online driving license registration and renewal etc [103].

Country	E-gov. development index		World e-gov. development ranking	
	2012	2010	2012	2010
Mexico	0.6240	0.5150	55	56
Panama	0.5733	0.4619	66	79
El Salvador	0.5513	0.4700	74	73
Costa Rica	0.5397	0.4749	77	71
Guatemala	0.4390	0.3937	112	112
Honduras	0.4341	0.4065	117	107
Belize	0.3923	0.3513	124	120
Nicaragua	0.3621	0.3630	130	118
Sub Regional Average	**0.4895**	**0.4295**		
World Average	**0.4882**	**0.4406**		

Figure 3.16 E Government Development in Central America

(SOURCE: United Nation E Governance Development Database 2012)

3.3.3 Barbados

- Edutech is an E Government portal facilitates the needs of education and supports E-learning [104].
- IJIS is a judicial portal that provides services regarding judiciary, criminal's records, laws information etc [105].
- ASYCUDA [106]

3.3.4 Central America

As per Figure 3.16, Mexico, Panama and EI Salvador are having good E Governance facilities and focusing online citizen participation using website as well as mobile applications [107].

It is clear from the Figure that Nicaragua, Costa Rica, Honduras, Belize are having poor ranking but still initiating E Governance projects to meet the current demand of stakeholders The following are the recognized E Governance project of various developing countries in Central America:

3.3.5 Mexico

- Mexico online projects allows public to effectively communicate with the government [108].

Country	E-gov. development index		World e-gov. development ranking	
	2012	2010	2012	2010
Chile	0.6769	0.6014	39	34
Colombia	0.6572	0.6125	43	31
Uruguay	0.6315	0.5848	50	36
Argentina	0.6228	0.5467	56	48
Brazil	0.6167	0.5006	59	61
Venezuela	0.5585	0.4774	71	70
Peru	0.5230	0.4923	82	63
Ecuador	0.4869	0.4322	102	95
Paraguay	0.4802	0.4243	104	101
Bolivia (Plurinational State of)	0.4658	0.4280	106	98
Guyana	0.4549	0.4140	109	106
Suriname	0.4344	0.3283	116	127
Sub Regional Average	**0.5507**	**0.4869**		
World Average	**0.4882**	**0.4406**		

Figure 3.17 E Government Development in South America

(SOURCE: United Nation E Governance Development Database 2012)

3.3.6 Panama

- Primera Dama website provides support to working women and poor people of rural areas with the collaboration of Ministry of Agriculture Development [109].

3.3.7 Guatemala

- E procurement Website provides online information regarding state procurement system.

3.3.8 Southern America

There are 12 countries belonging to Southern America in which Chile maintains its first rank with 0.6769 EGDI as per Figure 3.17. Colombia and Uruguay also have 2nd and 3rd rank respectively [110]. Established E Governance projects in some developing countries in Southern America are mentioned below:

Country	E-gov. development index		World e-gov. development ranking	
	2012	2010	2012	2010
United States	0.8687	0.8510	5	2
Canada	0.8430	0.8448	11	3
Sub Regional Average	0.8559	0.8479		
World Average	0.4882	0.4406		

Figure 3.18 E Government Development in North America

(SOURCE: United Nation E Governance Development Database 2012)

3.3.9 Ecuador

- Defensesodelpueblo.org.ec is a government web portal that provides administrative services to its citizen and also support online access to business related information [111].

3.3.10 Peru

- Information System for providing services to rural development [112].

3.3.11 Brazil

- Brazil's Interlegis is the project that provides online access to various administrative services to the user [113].

3.3.12 Chile

- Chilean Tax System provides support for online payment of tax to its citizens [114].

3.3.13 Northern America

Canada and USA are among two world leader countries of Northern America since 2003 as highlighted in Figure 3.18. Both countries have efficient online citizen services and provide easy access to its citizens. Local, state and US government are associated with a portal through which all the relevant information

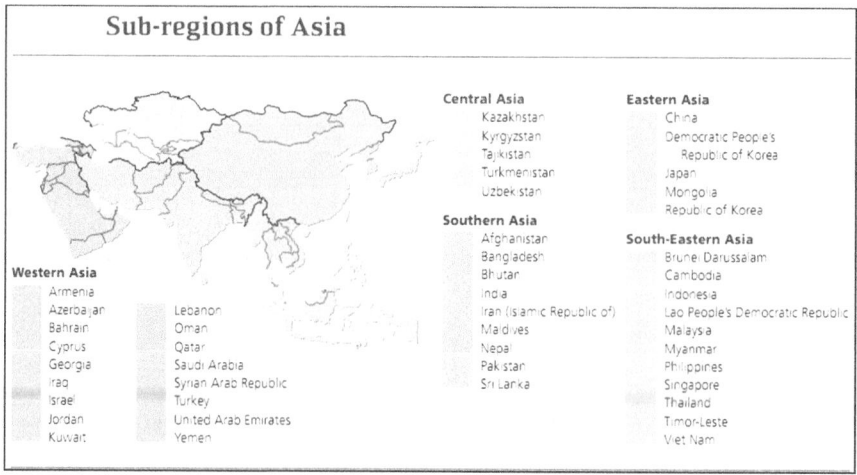

Figure 3.19 Sub Regions of Asia

(SOURCE: United Nation E Governance Development Database 2012)

can easily accessed by the citizens [115]. People are well aware of internet and day-to-day utility services are also managed by internet enabled user interface. Established E Governance projects in some developing countries in Northern America are mentioned below:

3.3.14 Canada

- Network of Canada Business Service Centers [116].
- Domestic Violence Front End Project.
- Canadian Consumer Affairs Gateway [117].

3.3.15 USA

- Social Security Administration provides security to the public and support easy and secure access to its services [118].

3.4 Status of E Governance in Asia

Asia is divided into five sub regions as indicated in Figure 3.19. Out of all Asian Countries Republic of Korea is leading and representing the Asia in the world with 0.9283 EGDI. Singapore, Israel and Japan also maintain their

Country	E-gov. development index		World e-gov. development ranking	
	2012	2010	2012	2010
Kazakhstan	0.6844	0.5578	38	46
Uzbekistan	0.5099	0.4498	91	87
Kyrgyzstan	0.4879	0.4417	99	91
Tajikistan	0.4069	0.3477	122	122
Turkmenistan	0.3813	0.3226	126	130
Sub Regional Average	0.4941	0.4239		
World Average	0.4882	0.4406		

Figure 3.20 E Government Development in Central Asia

(SOURCE: United Nation E Governance Development Database 2012)

rank in global scenario. United Arab Emirates has achieved rapid growth that improves its ranking in the world. It achieves 5th position in Asia and 28th in the world [119]. Norway and Emirates also facilitate the online services to its citizens. Region wise details of Asian countries are explained in subsequent section:

3.4.1 Central Asia

There are five countries in Central Asia although they are not resource rich but still doing fair in E Governance infrastructure developments. All the countries including Kazakhstan have developed their E Governance policy framework for enhancement of government services [120]. The rank of Kazakhstan is 1st in Central Asia with 0.6844 EGDI as shown in Figure 3.20:

The following are the recognized project in the E Governance of various developing countries in Central Asia:

3.4.2 Kazakhstan

- E Government portal for Kazakhstan provides user friendly services to its citizens [121].
- E pay website provides services where public can easily make online payment and E license website allow obtaining license in order to conduct particular type of services such as in transport, health, education etc [122].

Country	E-gov. development index		World e-gov. development ranking	
	2012	2010	2012	2010
Republic of Korea	0.9283	0.8785	1	1
Japan	0.8019	0.7152	18	17
Mongolia	0.5443	0.5243	76	53
China	0.5359	0.4700	78	72
Dem. People's Rep. of Korea	0.3616	N/A	130	N/A
Sub Regional Average	**0.6344**	**0.6470**		
World Average	**0.4882**	**0.4406**		

Figure 3.21 E-Government Development in East Asia

(SOURCE: United Nation E Governance Development Database 2012)

3.4.3 Uzbekistan

- E Governance portal for Uzbekistan government provide various E Governance services such as E Digital signature, E Taxation, improving Parliament capacity using ICT etc [123].

3.4.4 Tajikistan

- Border Management in Northern Afghanistan (BOMNAF) [124]
- BOMCA (Border Management Program in Central Asia) [125]
- An E Governance initiative for the anti-corruption campaign.

3.4.5 East Asia

The Figure 3.21 represents the 1st rank of Republic of Korea in the sub region of East Asia. Most of the official services provided by this country is online that is about 87 %. So it has achieved fastest growth and significant development in E Governance [126]. Japan has also secured 2nd rank in East Asia sub region and 18th position in the world E Governance ranking. China is also performing well in the E Governance sector. Providing and managing online services to a large population is a great achievement for China.

Established E Governance projects in some developing countries in Eastern Asia are mentioned below:

3.4.6 Republic of Korea

- KIPONET provides all intellectual property services through internet [127].
- HOMETAX is a system that allows online monitoring of the tax related issues such as collecting, deducting, and imposing and exempting without going to tax office and it is time saving [128].
- KONEPS manages all procurement procedure through internet [129].
- MINWON24 provide anytime, anywhere access of various civil services to the people via internet [130].

3.4.7 Japan

- E Government portal of Japan provides user friendly interface to its citizens to gain access to the legal statistics of Japan [131].

3.4.8 China

- E Government portal of China provides broad information, improving the transparency and allows easy access to different services of various sectors and also provides communication facilities between public and government [132].

3.4.9 South Asia

In the sub- region of South Asia, Maldives is at the top position and Iran is at 2nd position with 0.4994 and 0.4876 EGDI respectively as mentioned in Figure 3.22. Maldives and Iran are also performing well by providing efficient E Governance service to their citizen and securing good global ranking. Maldives has automated various government services such as vehicle registration and licenses etc. by providing a portal to its citizens [133]. Iran supports E Governance services by providing accessibility of government interfaces in English and Persian Language. India has secured 4th rank in South Asia and also progressing well in E Governance sector.

Established E Governance projects in developing countries of South Asia are mentioned below:

3.4.10 Pakistan

- NADRA provides e-passport facilities to the citizens [134].

Country	E-gov. development index		World e-gov. development ranking	
	2012	2010	2012	2010
Maldives	0.4994	0.4392	95	92
Iran (Islamic Republic of)	0.4876	0.4234	100	102
Sri Lanka	0.4357	0.3995	115	111
India	0.3829	0.3567	125	119
Bangladesh	0.2991	0.3028	150	134
Bhutan	0.2942	0.2598	152	152
Pakistan	0.2823	0.2755	156	146
Nepal	0.2664	0.2568	164	153
Afghanistan	0.1701	0.2098	184	168
Sub Regional Average	**0.3464**	**0.3248**		
World Average	**0.4882**	**0.4406**		

Figure 3.22 E-Government Development in South Asia

(SOURCE: United Nation E Governance Development Database 2012)

- Web based interface for various department such as finance, planning, establishment, health, interior, population welfare, housing and works [135].
- E Governance projects for skill development especially for small and medium scale enterprises [136].

3.4.11 Afghanistan

- Unique Identification number for citizens [137].
- E-Governance Facilitation Center [138].
- A Web based interface for providing interoperability among various department [139].
- ICT Village.

3.4.12 Sri Lanka

- Zainudeen. e-gov is a bulletin of E Governance published by Sri Lankan government focused to spread awareness regarding E Governance services available to the citizen for their convince [140].
- eSri Lanka Initiative (eSL) is also provides a transparent review about the functioning of government [141].

Country	E-gov. development index		World e-gov. development ranking	
	2012	2010	2012	2010
Singapore	0.8474	0.7476	10	11
Malaysia	0.6703	0.6101	40	32
Brunei Darussalam	0.6250	0.4796	54	68
Viet Nam	0.5217	0.4454	83	90
Philippines	0.5130	0.4637	88	78
Thailand	0.5093	0.4653	92	76
Indonesia	0.4949	0.4026	97	109
Lao People's Dem. Rep.	0.2935	0.2637	153	151
Cambodia	0.2902	0.2878	155	140
Myanmar	0.2703	0.2818	160	141
Timor-Leste	0.2365	0.2273	170	162
Sub Regional Average	**0.4793**	**0.4250**		
World Average	**0.4882**	**0.4406**		

Figure 3.23 E-Government Development in South East Asia

(SOURCE: United Nation E Governance Development Database 2012)

3.4.13 Bangladesh

- BanglaGovNet is a network of all government departments in district level as well as state level along with their centralized data center for various online transaction of government functioning [142].

3.4.14 South East Asia

There are 11 countries included in the survey and it is clearly established that Singapore is performing well among all countries of South East Asia region as indicated in Figure 3.23. It is one of the most innovative and well developed countries in the world.

Malaysia also maintains its 2nd rank in this region by providing attractive services to its people at various levels [143]. Malaysia also developed web based E health care system for achieving good care of health for its citizens. Established E Governance projects in some developing countries in South East Asia are described below:

3.4.15 Singapore

- OASIS system that collects and integrates different services relevant to various departments [144].
- Singapore Legal Policy Division [145].
- TrsutSg project [146].

3.4.16 Malaysia

- National Productivity Corporation is an E Government portal for government of Malaysia that offers online services to the public [147].

3.4.17 Thailand

- Thailand is having variety of E Governance initiative in which Mahathai.com is an important web portal for various online services required by citizens [148].
- Thai e-Passport is a quicker and automated way to complete the passport process in secure and authentic manner [149].
- Thailand is managing all taxation related service through its Online Car Tax Payment services [150].

3.4.18 Vietnam

- e-Visa Vietnam is an E Governance project for visa procedure. Vietnam having the facility of visa on arrival so all visa processes are controlled by automated software which keep tracks of all arrivals and subsequent departures [151].

3.4.19 Philippines

- COMELEC Modernization for voter that provides online registration of voter verifies their votes and support online counting of votes [152].

3.4.20 West Asia

Figure 3.24 shows the development of E Governance in West Asia. According to this figure Israel has secured 16th position as per global ranking and also maintaining various E Governance applications to provides easy and secure access to its citizens [153]. Saudi Arabia is also promoting E Governance application development especially in public sectors. A payment portal to handle online payments is also maintained by the government of Saudi Arabia.

Country	E-gov. development index		World e-gov. development ranking	
	2012	2010	2012	2010
Israel	0.8100	0.6552	16	26
United Arab Emirates	0.7344	0.5349	28	49
Bahrain	0.6946	0.7363	36	13
Saudi Arabia	0.6658	0.5142	41	58
Cyprus	0.6508	0.5705	45	42
Qatar	0.6405	0.4928	48	62
Kuwait	0.5960	0.5290	63	50
Oman	0.5944	0.4576	64	82
Georgia	0.5563	0.4248	72	100
Turkey	0.5281	0.4780	80	69
Lebanon	0.5139	0.4388	87	93
Armenia	0.4997	0.4025	94	110
Azerbaijan	0.4984	0.4571	96	83
Jordan	0.4884	0.5278	98	51
Syrian Arab Republic	0.3705	0.3103	128	133
Iraq	0.3409	0.2996	137	136
Yemen	0.2472	0.2154	167	164
Sub Regional Average	0.5547	0.4732		
World Average	0.4882	0.4406		

Figure 3.24 E-Government Development in West Asia

(SOURCE: United Nation E Governance Development Database 2012)

Established E Governance projects in some developing countries in West Asia are mentioned below:

3.4.21 Oman

- Oman Muscat Municipality website offers awareness regarding to the cleanliness of Muscat [154].

3.4.22 Lebanon

- Lebanon Education Center supports online education, training program and research for quality education [155].

3.4.23 United Arab Emirates

- E Government Municipal Services provides various online services [156].

3.4.24 Israel

- MyGov is an E Government portal for the Israel government and offers online access to various government policies, allows make online payments, access to administrative services etc [157].

3.4.25 Saudi Arab

- E Dash board portal provides online verification of citizen identity [158].

3.4.26 Qatar

- Hukoomi is the government gateway that collects and integrates different government services and provides access to these services via internet and mobile [159].

3.4.27 Iraq

- e-Iraq is an E Government portal dedicated to qualitative and fast public service delivery, designed on the basis of citizen centric model and contributing to establish good governance [160].
- Iraqi e-GIF is an E Governance projects support interoperability among various government departments.

3.5 Status of E Governance in Europe

The growth of E Governance in Europe is contributing 50% of the E Governance development of the whole world [161]. Whole European countries are partitioned into 4 geographical sub-regions as shown in Figure 3.25:

Western and Northern Europe provide better E Governance services as compared to Southern and Eastern Europe as indicated in 2012 survey. These countries are using common framework for E Governance for quick decision making and better citizen participation. European countries contributing

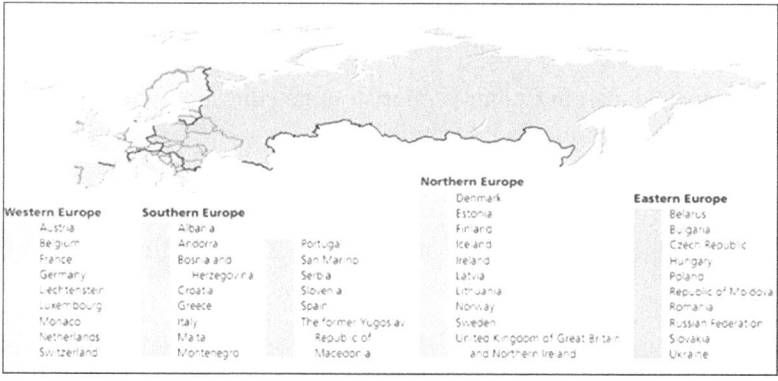

Figure 3.25 Sub-regions of Europe
(SOURCE: United Nation E Governance Development Database 2012)

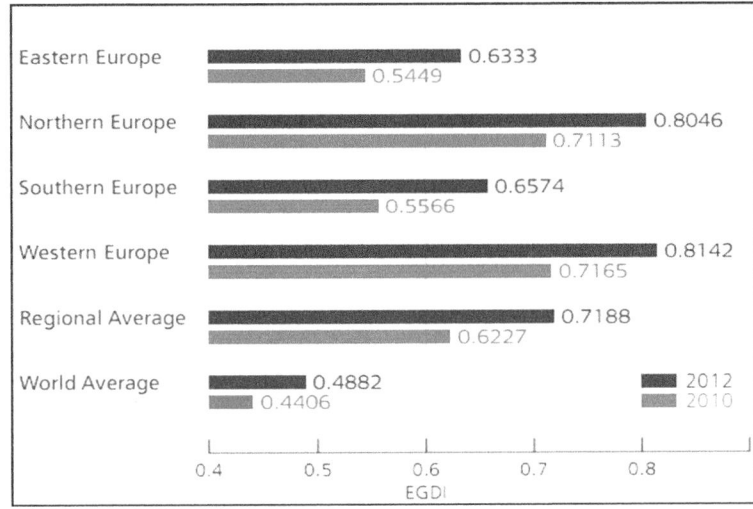

Figure 3.26 Regional E Government in Europe
(SOURCE: United Nation E Governance Development Database 2012)

towards the enhancement of E Governance services and by investing double of their per capita income for infrastructure development requirements of E-Governance.

3.5.1 Northern Europe

According to Figure 3.27, United Kingdom has secured 1st rank in the E Governance services with 0.8960 EGDI in Europe sub-region. All European countries have developed excellent E Governance infrastructure and ranked very close to each other [162]. All these countries are following common framework and focusing on citizen welfare through user centric E Governance model.

Following are some recognized projects of E Governance in Northern Europe Countries:

3.5.2 United Kingdom

- NHS Choices website provides services to millions of people and allows to access the health related information online. Healthpsace is an application that can be accessed with the help of this website. This application provides secure account to both doctors and patients. They can store, update and access their health data online [163].

Country	E-gov. development index		World e-gov. development ranking	
	2012	2010	2012	2010
United Kingdom	0.8960	0.8147	3	4
Denmark	0.8889	0.7872	4	7
Sweden	0.8599	0.7474	7	12
Norway	0.8593	0.8020	8	6
Finland	0.8505	0.6967	9	19
Estonia	0.7987	0.6965	20	20
Iceland	0.7835	0.6697	22	22
Lithuania	0.7333	0.6295	29	28
Ireland	0.7149	0.6866	34	21
Latvia	0.6604	0.5826	42	37
Sub Regional Average	**0.8046**	**0.7113**		
World Average	**0.4882**	**0.4406**		

Figure 3.27 E Government Development in Northern Europe

(SOURCE: United Nation E Governance Development Database 2012)

- Tell us once service provides information regarding birth and death details of citizens and it is responsible for transmitting significant information to the departments as per requirements [164].

3.5.3 Denmark

- EasyLog provides support to access the administrative services efficiently [165].
- NemHandel offers various services relevant to business infrastructure [166].

3.5.4 Estonia

- E Government portal provide access to various administrative services via internet [167].
- Health Information system provides services regarding health. It supports easy, fast and secure access to medical information to the Patients and Doctors [168].
- X-Road middleware [169].
- Public procurement state registers [170].

3.5.5 Lithuania

- Portal for Central Public Procurement [171].
- Secure state Data Communication Network facilitates the secure exchange of data over network among different parties.
- Electronic ID card provides the identity to the public by storing their signature and other relevant information in digital form [172].

3.5.6 Ireland

- E Tenders is an E Government portal and offers procurement services via internet [173].
- South Dublin digital books.
- Online Payment of Motor Tax [174].
- Revenue Online Service [175].
- Certificates.ie website enables the public to book and make payment of various certificates such as birth, marriage, death etc. [176].
- BASIS is an E Government portal that provides access to administrative services to the citizens. [177]

3.5.7 Eastern Europe

There are 10 countries participated in the survey and it is identified that Russian Federation (0.7345) has achieved highest rank among all countries in the Eastern Europe sub region [178] as shown in Figure 3.28.

Established E Governance projects in Eastern Europe countries are mentioned below:

3.5.8 Hungary

- ClientGate [179].
- Magyarorszag.hu.
- Electronic Government Backbone [180].

3.5.9 Czech Republic

- Public information portal provides online information about administrative services [181].
- E-Justice portal [182].
- Tax portal for making payment of taxes online without going to tax office [183].

Country	E-gov. development index		World e-gov. development ranking	
	2012	2010	2012	2010
Russian Federation	0.7345	0.5136	27	59
Hungary	0.7201	0.6315	31	27
Czech Republic	0.6491	0.6060	46	33
Poland	0.6441	0.5582	47	45
Slovakia	0.6292	0.5639	53	43
Bulgaria	0.6132	0.5590	60	44
Belarus	0.6090	0.4900	61	64
Romania	0.6060	0.5479	62	47
Ukraine	0.5653	0.5181	68	54
Republic of Moldova	0.5626	0.4611	69	80
Sub Regional Average	**0.6333**	**0.5449**		
World Average	**0.4882**	**0.4406**		

Figure 3.28 E Government Development in Eastern Europe
(SOURCE: United Nation E Governance Development Database 2012)

3.5.10 Bulgaria

- Egov.bg is the e- government portal that is based on internet and offering various services [184].
- eID cards.
- National Health portal [185].
- E-payment gateway [186].

3.5.11 Romania

- E Government portal of Romania provides access to procurement procedures and also maintaining transparency [187].

3.5.12 Southern Europe

In Southern Europe 14 countries have been considered for analysis of E Governance status. Spain, Italy and Greece are the important courtiers and progressing well in terms of E Governance infrastructure development. Although Spain slipped down in its world E Governance development ranking but it has secured its first position in Southern Europe regions. Other countries such as Slovenia and Croatia improved their EGDI ranking and secured 2nd and 3rd rank among sub-region countries of Southern Europe. It is also identified that native languages are widely used in E Governance applications especially in G2C interfaces [188]. The digitization of government data is prime priority of E Governance initiative in this region. The status of all countries in the sub region of Southern Europe is shown in Figure 3.29:

Some of the developed E Governance projects in Southern Europe countries are described below

3.5.13 Spain

- DNIe is an E Governance application in Spain containing huge database regarding citizen information, insurance information, financial information and information about all utility services for the citizen of Spain [189].

3.5.14 Slovenia

- E-Uprava system provides information related to the public administration services to its visitors [190].
- The aim of E-SJU portal to store all the services regarding administration in electronic form [191].

3.5.15 Italy

- MEPA is an E procurement portal that offers procurement services to its citizens [192].
- Magellano system allows the online payment of tax [193].

3.5.16 Portugal

- Public internet spaces system offers free access to the computer for all citizens [194].
- PORBASE is a database that maintains 1, 300, 00 bibliographic records [195].
- CITIUS is a system that allows online submission of court related documents [196].

Country	E-gov. development index		World e-gov. development ranking	
	2012	2010	2012	2010
Spain	0.7770	0.7516	23	9
Slovenia	0.7492	0.6243	25	29
Croatia	0.7328	0.5858	30	35
Italy	0.7190	0.5800	32	38
Portugal	0.7165	0.5787	33	39
Malta	0.7131	0.6129	35	30
Greece	0.6872	0.5708	37	41
Serbia	0.6312	0.4585	51	81
San Marino	0.6305	N/A	52	N/A
Montenegro	0.6218	0.5101	57	60
Andorra	0.6172	0.5148	58	57
The former Yugoslav Rep. of Macedonia	0.5587	0.5261	70	52
Bosnia and Herzegovina	0.5328	0.4698	79	74
Albania	0.5161	0.4519	86	85
Sub Regional Average	**0.6574**	**0.5566**		
World Average	**0.4882**	**0.4406**		

Figure 3.29 E Government Development in Southern Europe

(SOURCE: United Nation E Governance Development Database 2012)

- Electronic Vote project of Portugal stores voters details electronically and facilitating them to give their vote over the internet while they are not present in their designated voted area [197].

3.5.17 Malta

- E Identity system that store the personal information regarding citizen in electronic form [198].
- Online payment system [199].

- E Health portal system maintains the significant data regarding the public health and also provides secure and easy access to it [200].
- Online Customer care system.
- Judicial portal provides services regarding judiciary, criminal's records, laws information etc [201].
- Internet Phone Box service.
- E Tourism is a national portal that offers online service to its tourist [202].

3.5.18 Greece

- HERMS is a system that facilitates the public administration services [203].
- Taxisnet is a system that allows online payment of taxes without going to tax office and save time and also provides custom services [204].
- SYZEFXIS [205].
- GRNET [206].

3.5.19 Western Europe

As indicated in Figure 3.30, Netherland has secured 1st rank in the sub region of Western Europe by providing excellent online services to its citizens. It maintains a central database to collect and integrate the information from all departments and support easy data sharing and facilitate E Governance services such as electronic payment of bills, user identification etc. [207]. Following are the E Governance projects in Western Europe:

3.5.20 France

- *Vitale* is a system that provides online social insurance [208].
- *Taxation* portal enables online monitoring of tax related issues, filing personal income etc [209].
- *Marches Publics* provide services regarding E Procurement [210].

3.5.21 Liechtenstein

- An E Government portal is used to provide administrative services to Liechtenstein citizens [211].

3.5.22 Switzerland

- sme.admin website provides information regarding Small and Medium Enterprises [212].

Country	E-gov. development index		World e-gov. development ranking	
	2012	2010	2012	2010
Netherlands	0.9125	0.8097	2	5
France	0.8635	0.7510	6	10
Liechtenstein	0.8264	0.6694	14	23
Switzerland	0.8134	0.7136	15	18
Germany	0.8079	0.7309	17	15
Luxembourg	0.8014	0.6672	19	25
Austria	0.7840	0.6679	21	24
Belgium	0.7718	0.7225	24	16
Monaco	0.7468	N/A	26	N/A
Sub Regional Average	**0.8142**	**0.7165**		
World Average	**0.4882**	**0.4406**		

Figure 3.30 E Government Development in Western Europe
(SOURCE: United Nation E Governance Development Database 2012)

- Simap.ch is an E Government portal that supports the services of E Procurement [213].

3.5.23 Germany

- De-mail system provides the safety in online exchange of electronic information among people and various organizations [214].
- D115 project provides access to public administration services by using 115 number as a single point of contact [215].
- Electronic identity is achieved using a microchip that stores the personal information such as fingerprints or facial images in electronic form and recognize the identity of person over internet.

3.5.24 Luxembourg

- E Government portal facilitates the needs of education [216].
- Portal allows the public procurement services.

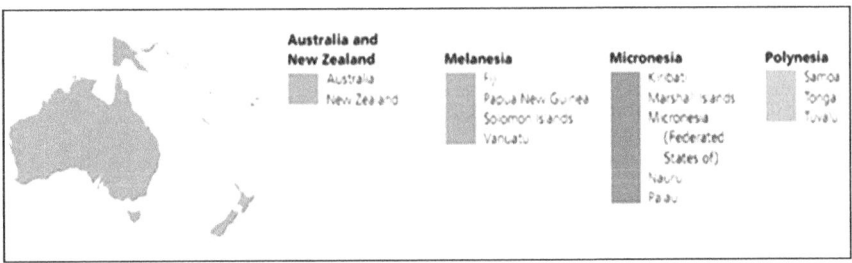

Figure 3.31 Region of Oceania

(SOURCE: United Nation E Governance Development Database 2012)

- Business Portal provides services related to business [217].
- Emergency Portal.
- eGO is a system that facilitates the e-payment services to Luxembourg public [218].
- Thematic portals provide access to the services regarding sports.

3.5.25 Belgium

- E-id cards are used for checking personal credentials [219].
- The national register contains some basic information of Belgian citizens.
- Crossroads Bank for Enterprise includes the source information for the authenticity of Belgian Enterprises [220].
- Crossroads Bank for Social Security Register contains the data regarding people who have registered themselves with social security of Belgium [221].

3.6 Status of E Governance in Oceania

As indicated in Figure 3.31, the continent of Oceania is divided into 4 sub regions for the monitoring of E Governance.

There are 14 countries in Oceania region in which Australia and New Zealand has achieved excellent performance in the development of E Governance practices as highlighted in Figure 3.32. Australia monitors the citizens' information through 900 websites of E Governance and provides online solutions to its citizen. These websites also contain the information of people and also facilitates the services such as online tax payment, monitoring the land and vehicle registration [222]. New Zealand is also providing E Governance

Country	E-gov. development index		World e-gov. development ranking	
	2012	2010	2012	2010
Australia	0.8390	0.7863	12	8
New Zealand	0.8381	0.7311	13	14
Fiji	0.4672	0.3925	105	113
Tonga	0.4405	0.3697	111	116
Palau	0.4359	0.4189	113	103
Samoa	0.4358	0.3742	114	115
Micronesia (Federated States of)	0.3812	N/A	127	N/A
Tuvalu	0.3539	N/A	134	N/A
Vanuatu	0.3512	0.2521	135	155
Nauru	0.3242	N/A	141	N/A
Marshall Islands	0.3129	N/A	146	N/A
Kiribati	0.2998	N/A	149	N/A
Solomon Islands	0.2416	0.2445	168	156
Papua New Guinea	0.2147	0.2043	177	171
Sub Regional Average	**0.4240**	**0.4193**		
World Average	**0.4882**	**0.4406**		

Figure 3.32 E Governance Status in Oceania

(SOURCE: United Nation E Governance Development Database 2012)

services to its people through single window Government interface for easy and secure access.

Established E Governance projects in Oceania region are mentioned below:

3.6.1 Australia

- E Australian Public Service Commission allows the implementation of government policies and provides easy and secure services to the public [223].

3.6.2 New Zealand

- *Government Shared Network* allows sharing of information at rapid speed among all government organization in cost effective manner. It also provide user friendly interface to the citizen for information delivery [224].
- psi.govt.nz is an intranet service developed on the model of single window application maintains a collaborative network of all government officials for information exchange [225].

3.6.3 Fiji

- The main E Governance initiative in Fiji is e-community center which provides internet accessibility in all interior places of Fiji. It also encourages citizen participation and dedicated to bridge the gap between computer literate and illiterate. It is also hosting website for E Learning and E Health for E Governance purpose [226].

3.7 Conclusion

In this chapter worldwide status of E Governance initiatives have been discussed. It has been clearly established that now there is a need to interconnect all departments of the government so that a common E Governance system could be utilized for better citizen centric services. This integration enhances organizational collaboration thus improvement over efficiency, effectiveness and citizen services. It is also found that most of the government organizations are launching their official websites. Although it is a beginning step but sufficient enough to ensure transparency and easy access to the citizen. It is also challenging to update the websites continuously so that the citizen could able to rely on the organization.

It is observed in the survey that least developed regions especially few African regions are lacking skilled resources, infrastructure as well as development plan so there is poor governance with substantial scope of improvement. Other least Development countries such as Mali, Afghanistan and Bhutan are still limited to only website development as their E Governance initiative because of poor infrastructure, limited fund and unskilled manpower.

It is apparent from the various facts that all European countries are doing well in E Governance because a common policy framework is followed by all these countries and they are more politically stable. The world wide status of E Governance is showing varying pattern because all regions are entirely different in terms of their social, political, environmental and financial constrains. It is very challenging to indentify a common policy framework which can serve the whole world but it is extremely essential to adopt "Think Global Act Local" policy so that all efforts could join to bring a new society and a bright future to the mankind.

4

Status of E Governance in India

4.1 Introduction

In India E Governance is a revolutionary deployment of computerization in Governmental Departments to achieve citizen centric and transparent governance. Around this ideology, a huge country wide infrastructure has been developed and large-scale digitization of data is taking place to endow the seamless data access and connectivity. The ultimate objective is to provide easily accessible government services to the rural as well as urban people. The nationwide E Governance Plan (NeGP), takes a universal outlook of E Governance across the country by incorporating into a collective vision as well as distributed determinants [227]. The National E Governance Plan by the Government of India prepared the groundwork of E Governance and offers long-term revolutionary growth in governance applications. The sole aim is to initiate various projects at central, state and local level to create a citizen centric environment for good governance. Figure 4.1 shows the importance of E Governance.

4.2 Categorization of E Governance Projects in Multi-Tier Mission Mode Projects (MMPs)

Mission Mode Projects (MMPs) are individual projects assigned in the National E Governance Plan (NeGP) which focuses on a certain feature of E Governance, such as in banking, property records or income taxes etc. NeGP consists 31 Mission Mode Projects (MMPs) which are further classified in 3 levels of architecture: state, central or integrated projects [228]. Each state government can also define five MMPs itself. MMPs are controlled by various Ministries as:

- Central.
- State.
- Integrated.

E Governance Data Center, Data Warehousing and Data Mining: Vision to Realities, 73–94.
© 2013 *River Publishers. All rights reserved.*

Figure 4.1 The Importance of E Governance

4.3 Central Level Mission Mode Projects

The Figure 4.2 shows various government organizations which comes under Central Mission Mode Projects (MMPs) categories.

4.3.1 E Governance in Central Excise

The Central Board for Excise and Customs has adopted a software system to conduct their trade and service tax operations [229]. The aim is to facilitate better administration, transparent work flow, efficient monitoring and accountability. The objectives of the project are mentioned as follows:

- The key idea of this program is to re-engineer organizational procedures and offer tax management into a contemporary, well-organized and transparent manner.
- E Filing and E Processing of documents has started in the paperless manner and minimized various overhead involved.

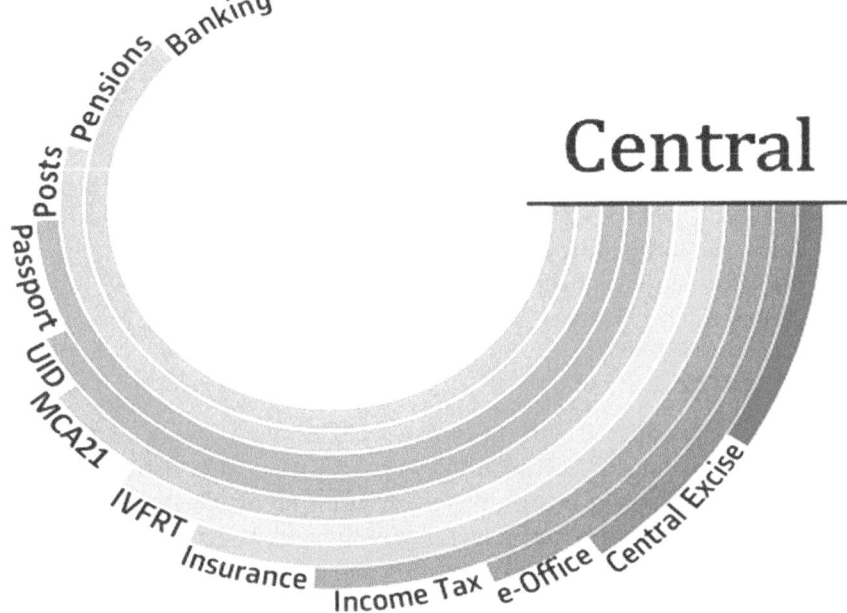

Figure 4.2 Governance Sectors Coming Under Central MMPs

4.3.2 E Office

The Government of India has taken initiatives to update the E Office Central Government Bureaus through the commencement of Information and Communication Technologies. The E Office Mission Mode Project has been taken up to improve efficiency in government working procedures and service distribution. The product is aimed to optimize the existing work flow, file routing, assessment and office instructions. It is also equipped with online reporting, web based forms and digital signatures for authentication [230]. The objectives of the E Office MMP are:

- Decrease the delay of processing.
- Establishment of the transparency.
- Decrease the turnaround time and satisfy the needs of citizen's charter.
- Provide utilization of resources to improve the administration quality.
- Improve the efficiency and consistency of government response.

4.3.3 Income Tax Department

The Income Tax Department of India has been developing an application for setting up a wide-ranging on line services regarding automation of the department with following features [231]:

- Country level consolidated Income Tax Application.
- National Level PAN Module.
- Electronic filing of Income Tax Returns.
- Tax Information Network (TIN).
- Refund Banker Scheme.

4.3.4 Insurance

The revolutionary change in the Insurance sector through private involvement has becoming technically advance and more up to date. It is also evident that the competitiveness is growing among Government/Non Government Public Insurance Companies. Private players are embracing latest technologies and fascinating new customers through new products, smart marketing and quick delivery.

In the last few years, a huge number of insurers, both life and nonlife, have recognized their presence in a central database of Insurance sector. Customer service is the trademark of this sector and it became crucial for the Public Sector Insurance Companies to automate their processes and control the contemporary technologies to provide a world class facility to customers [232].

4.3.5 Immigration, Visa and Foreigners Registration and Tracking

India is attractive for foreigners as an important tourist destination. Immigration authentication is the primary step which creates fair impression about the country, thus it is essential to have quick user-friendly services for all foreigner visitors [233]. An intelligent and secure framework is required to adopt which could efficiently integrate Ministries of Home as well as External Affairs including Central Board of Excise and Customs, and Civil Aviation during the Visa issuance process. The project is already started and currently it is under the process of renewal and up-gradation.

4.3.6 MCA21

MCA21 project is initiated by the Ministry of Corporate Affairs (MCA), the Government of India. It gives simple and secure access to services provided by

MCA for various entities such as corporate entities, professionals and public. The processes where the legal requirements are enforced and complied, fully automated by MCA21 project, under the Companies Act-1956 [234].

4.3.7 Aadhar Card

The project for unique identification of citizens was started to allot unique identification to each resident across the country. The Unique Identification Number would help to make better governance in different states particularly in social welfare related schemes [235]. It would significantly decrease the identity scams and help to locate the correct beneficiary. Over a period of time, this may help to decrease the total expenses under these schemes by avoiding the possibilities of duplicate identities under the same scheme.

The allotment process of unique identification number is shown in Figure 4.3. The concept of a unique ID was first discussed in 2006. On 3rd March, 2006 the "unique ID for BPL families" was given by the Department of Information Technology. The expected unique identification project users are described by using Figure 4.4.

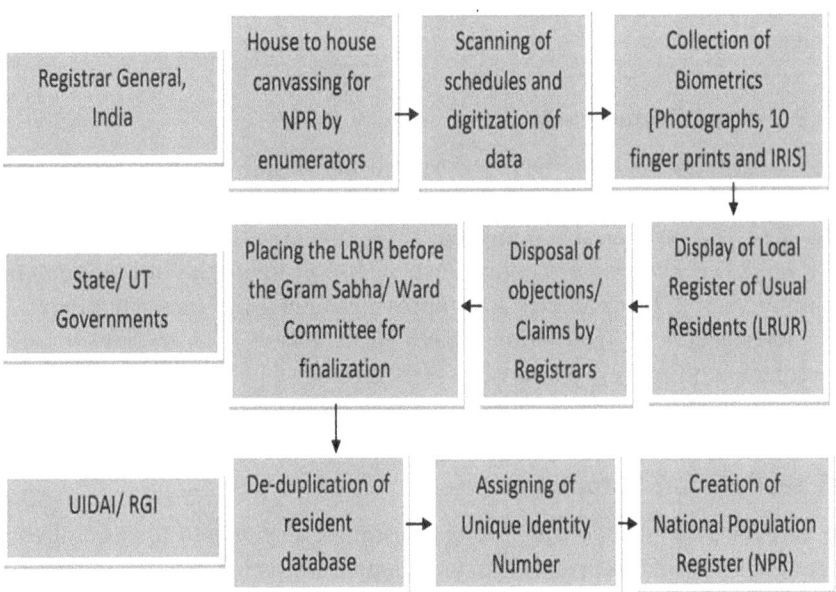

Figure 4.3 National Identity System

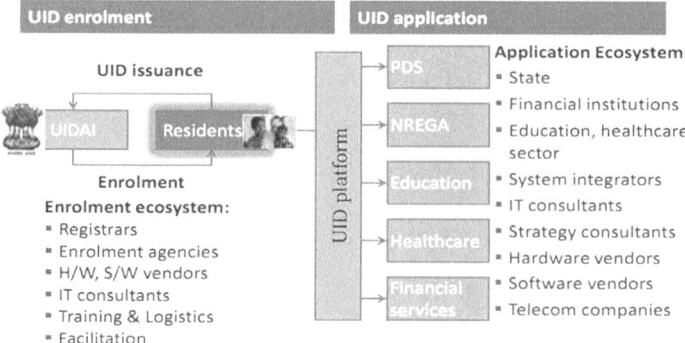

Figure 4.4 Importance of Unique Identification System

4.3.8 Passport Department

The Passport Division of the Ministry of External Affairs (MEA) has launched Central Passport Organization (CPO) for all passport related services [236]. It is a secure automated system which includes verification and validation process for each application under consideration. The ongoing project includes process re-engineering, latest technologies and highly reliable working environment so that the rising demands of passport related services could be discharged shortly.

4.3.9 Postal Department

The Department of Posts has modernized the postal services through networking and computerization of all Post offices. It is deployed by using a central server system and setting up of Computerized Registration Centers (CRCs) [237]. The Department of Posts is also utilizing the Information and Communication Technologies for improving money transfer operations and their interfaces with international payment gateways. The tracing of registered items and parcels etc. are possible through the official website of the department.

4.3.10 Pension Automation

An automated application regarding pension management has been already started. The aim of this project is to maintain a database of retired people along with their admissible retirement related benefits. It also has a grievances re-addressable system to bridge the gap between stakeholders and policy makers.

S. No.	Details of E Governance Projects	Controlling Authority
1	Kiosk	SBI
2	e-Sanjeevani	C-DAC Mohali
3	Language Technologies	Department of Electronics and Information Technology (DeitY)
4	Bharat Operating Systems Solutions (BOSS)	Department of Electronics and Information Technology (DeitY)
5	Multi-sectoral Development Programme (MsDP)	C-DAC Mumbai
6	Passport Seva Project,	Ministry of External Affairs
7	E-Nose, E-Tongue and E-Vision	C-DAC Kolkata
8	Best Practice E Governance Solution	National Informatics Center
9	Sanyog, Punarbhava portal,	Media Lab Asia, Delhi
10	Computerization of Land Records	Government of India

This application includes online availability of all orders/instructions on the portal for user access [238].

4.3.11 Nationalized Banking

The priorities of E Governance in Banking System are improved operational efficiency and speedy transactions. It has the objective of anytime, anywhere banking to Indian customers [239].

4.3.12 Other Central E Governance Projects

Following are some other E Governance project managed at Central level [229].

4.4   State Level Mission Mode Projects

There are 11 state MMPs which would be possessed and controlled by various State Governments as shown in Figure 4.5. Each state working in a project within a quantified timeframe is required to prepare a complete documentation either in-house or with the assistance of a consultant. The level of facilities and desirable features of the projects could be determined by consultation with the customers or stakeholders. State Governments are answerable for implementing the State MMPs under the leadership of the respective authorities [228] [240].

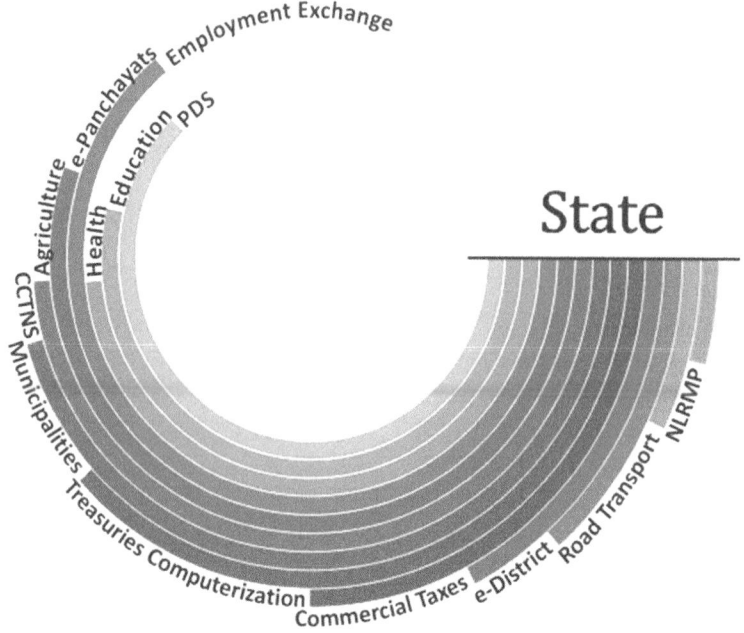

Figure 4.5 Governance Sectors Coming Under State MMPs

4.4.1 National Land Records Modernization Program

A Project for Automation of Land Records was launched in 1988-89 to eliminate the inherent faults of maintenance in the manual system. In 1997-98, the scheme was drawn-out to tehsils to allocate the records by the landowners [241]. The focus of the whole operation is to perform digitization of land records so that easy access of land record could be possible and it could also record the details of ownership over a period of time for maintaining transparency.

4.4.2 Road Transport

The Ministry of Road Transport and Highways has been enabling the procedure of automation of nearly one thousand offices across the whole country within last 5 years and currently 90% of the RTOs have been automated. Since the documents produced by the RTOs (Driving License) are legal across the country, it is essential to have a trustworthy system so that correct document could be produced. The Smart Card standards are already under development and VAHAN and SARATHI are the two software which have successfully implemented across the country [228] [242].

4.4.3 E District

In Indian administration system, most of the Government-to-Citizen communication takes place at District level. The E Districts project enhances the effectiveness of various district-level departments. Under this scheme, the front end is an interface for citizen facilitation centers and it is deployed at district, tehsil, sub division and block levels. Common service centers for delivery of services are used to establish Village-level front ends [228]. The Table 4.1 describes the services that are planned to be delivered through this MMP [243]:

4.4.4 Commercial Taxes

There is a consistent demand to streamline VAT administration by establishing standards with respect to Commercial Tax (CT) and citizen-centric, service oriented processes. In India, in a normal case, Commercial Tax departments accounts for 60-70 % of total revenue of States. This initiative is led by Department of Revenue (DoR), Ministry of Finance, along with strategic consultation by Ernst and Young (E&Y) and National Institute for Smart Government (NISG) [229] [244]. Some recommendations to facilitate the simplification of administrative procedures in Commercial Tax are given below:

Table 4.1 E District Services

Services	Description
Public Distribution System	Issue regarding Ration Card etc
Utility Payment	Payment regarding electricity bills, water bills etc
Assessment of Taxes	Government taxes such as property tax etc
Certificates	Certificates of Birth ,Domicile ,Caste ,Death ,Income etc
Licences	Driving license, arms license
Information Dissemination	This service is described about various government schemes.
Right To Information	Online filling and providing receipt of information regarding RTI act
Social welfare scheme	Disbursement of Family pensions, widow pensions and any other old age pension
Complaints	Complaints of non-availability of Doctor, absentee teacher etc.

- Provide various online information services to the dealer.
- Filing of returns in electronic form.
- Refunds Clearance electronically.
- Payment of taxes in electronic mode.
- Online issuance of various Central Sales Tax related forms.

4.4.5 Treasury Computerization

The State Treasuries are the organizational and financial unit of the Government Monetary System and are liable for handling the day-to-day trades of receipt and payment of Government [245]. The functions generally implemented by the treasuries are:

- All money collection of Government.
- All Payments of Government.
- Management of Retirement Fund.
- Sale of Stamps through Vendors.
- Compilation of Government Accounts (District wise).
- Safe custody for Treasures.

The objectives of the Treasury Computerization MMP are:

- More Efficient Budgets.
- Cash Flow Improvement.
- Real-Time Account Reconciliation.
- Strong MIS (Management Information Systems).
- Better accuracy and suitability in account preparation.
- Efficiency and transparency associated with public delivery systems.
- Better Financial Management.
- Improved governance in States and Union Territories.

4.4.6 Municipalities

The main purpose of Municipalities MMP is to provide efficient and effective municipal services to citizens. The goals of the MMP are given below [246]:

- Improve the productivity rate of Urban Local Bodies.
- Adoption of an integrated standardized approach.
- The management of an Information System for effective decision making.
- Single windows services to citizen, round the clock.

4.4.7 Crime and Criminal Tracking Network and Systems (CCTNS)

This project has been accepted by CCEA (Cabinet Committee on Economic Affairs) in June, 2009 with a budget of 2000 Crores. The key objectives of the CCTNS MMP are to [247]:

- Provide Improved Investigation tools, Crime Avoidance, Law and Order Preservation, Traffic Administration, Emergency Response, etc.
- Utilize IT for efficiency and effectiveness of core supervising operations.
- Provide information for easier and faster examination.
- Growth in Operational Efficiency by:
- Decreasing tedious and repetitive manual task.
- Automating back-office functions, and make sure that police staff doing their duties or not.
- Making a common platform for State and Central level to provide sharing of information or databases of criminal or crime across the nation or states from corner to corner of the country.
- Build a platform for intellect sharing across the country.

4.4.8 Agriculture

DAC (Department of Agriculture and Cooperation) has undertaken numerous Information Technology oriented solutions like AGMARKNET, SEEDNET, DACNET etc. Agriculture Mission Mode Project recommends to join in these IT initiatives with fresh applications / modules and being the part of the Project [248].

In order to attain the aforesaid vision the department has developed an online platform as shown in Figure 4.6 [227] [248]. This includes following key features:

- Single window solutions for the farmers, for accessing relevant information and services during the whole crop-cycle.
- Multiple channels for information.
- By decreasing the time between generating and broadcasting of information.

4.4.9 E Panchayats

Panchayat is the first level of government interaction for more than 60% citizens of India. It provides various services to millions of people of rural

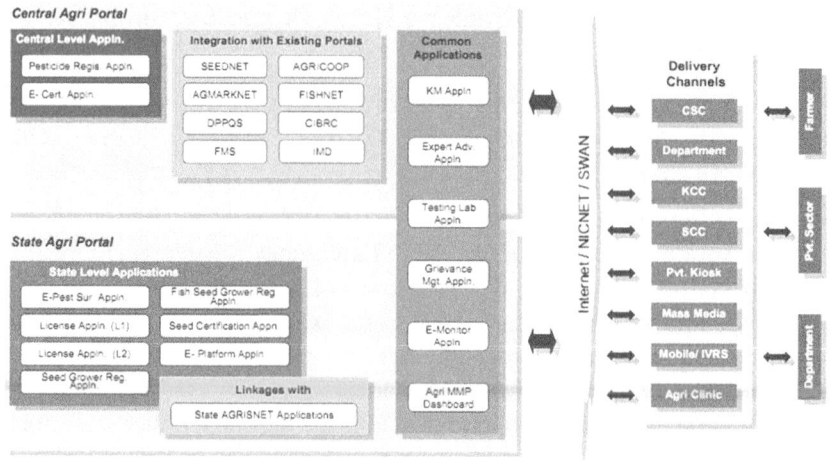

Figure 4.6 Accessing Services by an Online Platform

areas. The main purpose of the Panchayat MMP is to solve the main challenges faced by the rural community [228]. For example there is no proper infrastructure for reliable communication; people don't get service on time, problem with monitoring mechanisms for schemes and irregularities in utilization of fund [249]. The functions and services of Gram Panchayat MMP are:

- Solve the issue regarding certificates of Birth, Death and Income etc.
- Provide the receipt of Progress report, various funds etc.
- Solve issue of House-related services.
- Dissemination of BPL data.
- Issue of trade licenses and NOC.
- Provide proceedings' copy of Gram Sabha.
- Dissemination of internal processing of Panchayat Agenda.

4.4.10 Employment Exchange

This MMP is being conceptualized by the Ministry of Labour and Employment. The aim of this project is to match the needs of employer and job seekers. Important functionalities expected by this MMP are career guidance to unemployed and online vacancy registration by employers [227] [250].

4.4.11 Health

Ministry of Health and Family welfare is using Information and Communication Technologies (ICT) for management of various programs like Mother and Child Tracking System [228]. Ministry also aimed to use ICT through MMP, for benefits in various schemes like, Supply Chain Management for vaccines and medicines, Hospital Information Systems and National Rural Health Mission [251].

4.4.12 Education

Education in India requires an innovative working framework that includes active participation of Department and Ministries at central and state levels [227]. Extensive coordination is needed within government agencies for successful administration of mid day meals, hostels and disbursement of scholarships. ICT will be very instrumental in implementation and monitoring of flagship schemes like Mid Day Meal, Sarve Shiksha Abhiyan, Rashtriya Madhayamik Shiksha Abhiyan etc [252].

4.4.13 Public Distribution System

The Public Distribution System (PDS) has been included as a Mission Mode Project after approval with the competent authorities [228]. In this project, the key issues are Supply Chain Management with proper allocation and utilization reports, storage and movement of food grains, system for grievance redressal, digitisation of beneficiary database, Fair Price Shop automation and transparency portal etc [253].

4.4.14 Other State E Governance Projects

Table 4.2 shows some E Governance project managed at state level.

4.5 Integrated Mission Mode Projects

Following are the projects shown in Figure 4.7 come under the category of Integrated Mission Mode Projects [229][254]:

4.5.1 Common Services Centers (CSC)

CSC is an E Governance government to citizen interface gateway. It is used to cover E Governance services on a massive scale through its single window operation. CSC provides high quality services in the following areas [227]:

Table 4.2 State wise E Governance Project

States	Project Name
Andhra Pradesh	SUJAL-Waer Quality Monitoring System, Agiculture-AGROMET, M-Health, Vaccination Appointment for International Travelers Online Booking System, E-Hospital, Missing Persons, Agriculture-NHM, Online Drug-Supply chain management System, Pension Management Information System for State Govt. Employees, Online Admissions to Various Degrees in A.P., E-Panchayat for Panchayat Raj Dept.- Gram Panchayat Specific, e-Pancnayat for Panchayat Raj Dept. - Common to all PRIs, CARD - IG Registration and Stamps, Land Records, M-Foods, Adarsha Rythu
Arunachal Pradesh	Treasurynet, Gpf, Agmarknet, Jansuvidha
Assam	Land Records, Gram Panchayat, Election, NREGASoft
Bihar	Vahan-Sarathi, e-Gazette, DACNET, MCTS, e-Courts, e-Panchayat
Chandigarh	CSC SAMPARK, Jan Sampark, Land Records, GePNIC, Agnculture -agmarknet, Treasuries, Intranet Portal of Chandigarh Admimstration, Prevention of Food Adulteration Licences for Health Department, State Portal
Chhattisgarh	e-Panjeeyan, NREGA, Paddy Procurement - Online, Public Distribution System - Online, e-GramSuraj, Jandarshan, CGBSE, GAD Online, e-Rojgar, OSMS, RSBY, Daily Hospital reporting System, Online VET-MIS, Agri-MIS, Online Agri-Budget, Mandi Online, Agri-Subsidy, AGMARKNET, Chhattisgarh LokAyog Monitoring System, Health Program Monitoring, e-Mahatari, PHEDCMS, PRIASOFT, ERMS
Delhi	Web based Online counseling for Technical Ecucation, e-Pramanpatra, DORIS, e-SLA, DLRC, GPAMS, Computenzation at PFA , e-Awas, Pulse Polio Information System, e-Litigation, Centralized Payroll, GPF Information System, Marriage Information System, e-Purti, e-Yojana, e-Gazette, e-Clipping, VIR_SFU, PGRAMS, Letter Monitoring System
Goa	MAS, Infogram, Dharani, VATSoft, DC*Suite, Taluka*Suite, Vahan and Sarathi, XGN
Gujarat	e-Ration Card, District Passport Application Collection Centre, SWAGAT, City Survey Information System, e-Mamta, e-ITl Admission, Employment Exchange, Lane Records, GARVl, ATVT, OJAS, NOT MMP, XGN, DLIMS, XLN, Revenue Cases Management System, iOjN, Vikaspath, Online Recruitment process for Judicial officers

States	Project Name
Haryana	NLRMP, HALRIS, e-DISHA, RECORD Revenue, Hospital MIS, SJE, SARATHI, VAHAN, Harsamadhan, Centralized Grievances Redressal and Monitoring System, SRCbTPDS, Electricity Bills and Connection System, Elections Voters, Labour On-line Registration, CIPA, Arms License MIS, e-Posts. NICNET
Himachal Pradesh	Employment Exchange, RefNIC, ePraman /ePehchan, HIMPOL, Shastr, eSalary, Vahan, Online Hotels Reservation System, Sarthi, Agmarknet, HP OLTIS, Results Dissemination, eGazette, eKalyan, eCourts, HIMRIS, HimBhoomi
Jammu AND Kashmir	iTISP, VAHAN, SARATHI, CoIS, Immigration Check post Computerization, RuralSoft, CIPA, GePNIC, BIS, BDIS, AGMARKNET, eCourts, CONFONET, CCMS, FTS, NREGA, NSAP, DISE, SSA, GenProfit, CPIS, CPFIS, AWAZ-E-AWAM, Online Public Grievance Monitoring System
Jharkhand	vahan and Satathi, CEP, Prison Management and Visitor management system, Video Conferencing between Jails and Courts, e-Post Office
Karnataka	BBMP, Nemmadi, HRMS, Bangalore One, Karnataka One, E-PROC, KSWAN, e-High court of Karnataka and Circuit Benches, Karnataka Lokayukta, Smart Card issue of VAHAN and SARATHI, Scholarship Management Information System, CET, PLO System, e-tax, Bhoomi
Kerala	DC*Suite, RR online, SAND, EMS Housing Scheme, Pattayam, LINK, TRIM, Confonet, Agmrknet, Swabhiman, Voter SMS, CIPA, SEVANA, BLOOO DONOR'S Directory, e-Manal, RELIS, eHajar, Online Scholarship Management System, RCMS, Ayyankali, BPL, MGNREGA, IMD website Doordarshan website, AG office application, CAPnic, hsCAPnic, hsTransfer Trend-LB, Trend - LA/HP - LA/HP election, LCAPnic, B Tech Online, M Tech Online, EMLI, BOUGETTE
Lakshadweep	Vahan and Sarathi, CIPA, MGNREGA, COMPACT, Land Records, Fishnet, e-Courts, SSA
Madhya Pradesh	MP School Education Portal, IT1 Portal, e-Kanij, PARAKH, Samadhan Online, Samadhan Ek Din, e-Office, Citizen Services Management, Public Service Management System, Passport Control and Issuance System, BhuAbhilekh
Maharashtra	CIPA, CONFONET, MNREGA, NSAP - IGNOP, AGMARKNET, MPLADS, IDSP, CGHS, VC for CICs, DSC, RealCRAFT
Manipur	CIPA-Manipur Police, Payroll, e-Pension, VAHAN, SARTHI, e-Procurement in PMGSY, AGMARKNET, Results Dissemination
Meghalaya	e-district Services, VAT, GePNIC, TreasuryNET, MPSC, CONFONET Project, MNREGA, NSAP - IGNOP
Mizoram	Commercial Taxes, Agmarknet, passport, Agriculture

(Continued)

Table 4.2 Continued

States	Project Name
Nagaland	Personal Information System, Weather Information System, Web Services (State Departmental Websites), e-Court, Online Scholarship Management System, AGMARKNET, Arms License Computerization, Election Randomization, Electoral Roll Data Base, PMGSY, MGNREGA, BRGF, NSAP Transport (Sarathi and Vahan), MCTS - Mother and Child Tracking System, VAT Soft Computerization, CIPA, PHED-RGNDWM, DDWS, Minor Irrigation Census, NADRS, File Tracking System, State Official Website
Orissa	panchayat computerization Project, VATSoft, Vahan and Sarathi, e-Election, online Electricity bill pay
Punjab	Web based Online counseling for Technical Education, SUWIDHA, PRISM, ITISP, VAHAN, SARATHI, SSIS, AIS, CIPA, GePNIC, BIS(Budget Information System), ALIS, BDIS, MAREG, MARRCIS, AGMARKNET,E-Kiosks, CONFONET, CCMS, CoIS, BPIS, Rural Soft,DSMS, HCIS, FTS, NREGA, NSAP, MESS, eEntryPass, DISE, Visitor Management System, NADRS
Rajasthan	IFMS, LRC, NREGA-Soft, Immigration Control System, e-Exchange, PMGSY, Forest Land data Bank, e-Forest Right
Sikkim	NLRMP, NREGASoft, CPAO
Tamil Nadu	e-District, e-Social Welfare, e-Revenue, E-Procurement, e-Tendertng, Taxes e-Returns, e-payment, e-Request, E-Accounting, E Governance Portal for Directorate of Technical Education, e- Transport Booking, Nilam (Land Records), TNPCB, TNPSC
Tripura	Physically Challenged Personal Information System, e-Hospital, Onconet, Online Blood Bank Status and Bllod Donor Search, Energy Billing System, Land Records, Bhunaksha, NLRMP, e-Suvidha, Indo Bangladesh Passport Status Online, Tripura Information Commission (TIC) Online, e-Pourasabha, s-PSC, AAS, RuralSoft, Online Room Booking System
Uttar Pradesh	e-Ration Cards, E-Scholarship, Lokvani, Tehsil Divas, Bhu-Lekh, VAHAN and SARATHI, Nagar Soft, PRERNA, Niyukti On-line, Rural Soft, COSIS, PFAD, PlanPlus, Basic Shiksha, Pay-roll
UttaraKhand	MNIC, e-janadhar, Voter Search, MCTS/NRHM, State Web Portal, Computerization of UKPSC, Ganga-XGN
West Bengal	Land Records, Computerization of Registration of Property(CORD), E Governance in Commercial Tax, Transport (VAHAN and SARATHI) Project, Computerization of Indian Major Ports, eCourts, e-Procurement, Agriculture, Police, Employment Exchange, OBC and SC/ST Certificate Application and Review (OSCAR)

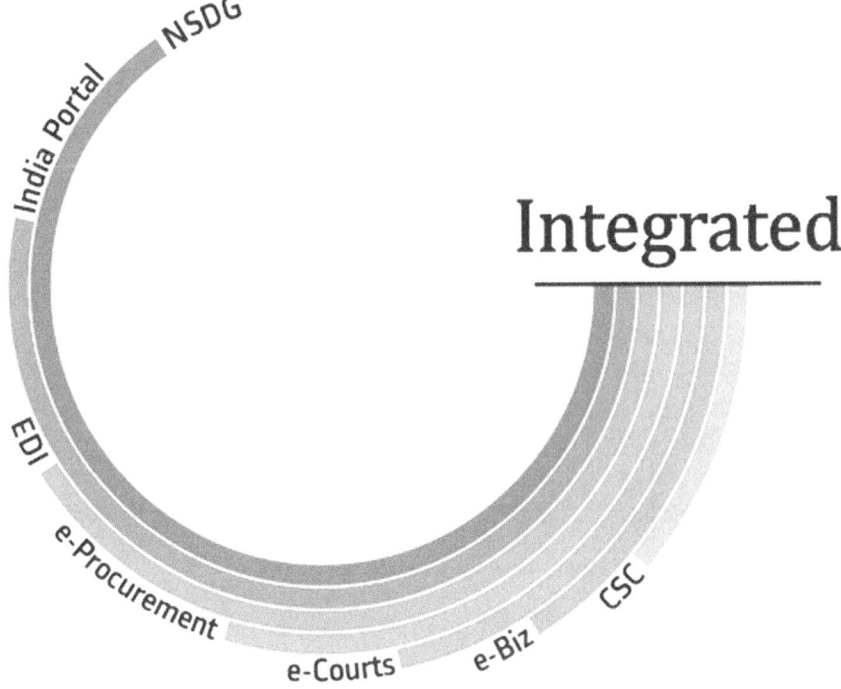

Figure 4.7 Governance Sectors Coming Under Central MMPs

- Education Portals.
- Telemedicine.
- Health Portals.
- Entertainment.
- E-governance.

CSC will provide various E Governance services such as issuance of certificates, public distribution system, and utility expenditure etc. trough a web enabled interface especially for rural areas. It also spread awareness and encourages NGOs and private sectors for active participation in various CSC schemes. CSC plays an active role for the development of rural areas [228][255]. The Public Private Partnership (PPP) model of CSC scheme has 3-tier architecture as described below:

- **CSC operator**: It is the first tier of PPP model. It is also called Village Level entrepreneur.

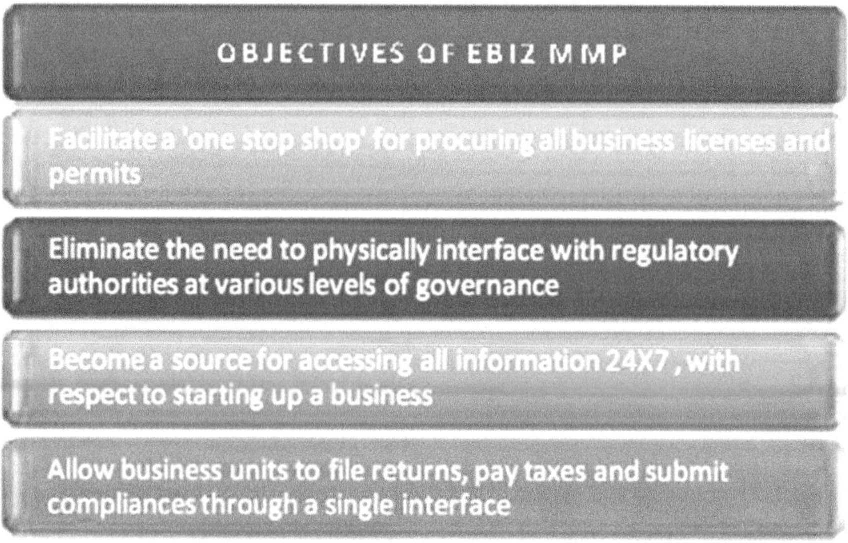

Figure 4.8 Objective of e-BIZ MMPs

- **Service Center Agency**: It will handle division of 1000 CSCs.
- **Nominated agency at State level**: It is a state level authority responsible for management of the MMP.

4.5.2 E Biz

The E Biz Assignment Mode Project [256], is initiated by the Ministry of Trade and Industry, is committed to provide efficient, convenient, clear, integrated electronic facilities to investors, industries and commercial organizations [228]. The objective of E Biz is shown in Figure 4.8

Like any other skill driven project, E Biz too has a clear order of addressing several issues for smooth and better government to industry interactions. It could easily facilitate following features:

- Value to Investors.
- One Stop Shop.
- Reduced Total Time and Cost.
- Anytime, Anywhere, Anyhow Government.
- Transparency and Secure Dealings.

4.5.3 E Courts

E Courts is the implementation of Information and Communication Technologies in Indian Judiciary. In this proposal, there are 3 phases in a period of 5 years. The basic goal of MMP is to automate the decision-making system and deliver it in 700 courts that are located in Delhi, Mumbai, Kolkata and Chennai, 900 courts situated in around 29 capital cities and 13,000 courts at district and subordinate level across the Nation [228][257].

The objectives of the E Court MMP project are [227]:

- Introduce transparency in the information.
- Provide easy access to official databases.
- Reducing the pendency of cases.
- Reform the day-to-day activities of judicial administration.

4.5.4 E Procurement

The purpose of this project is to make clear, simple and result-oriented government procurement process. It is a web based applications which manages optimal delivery of public amenities, efficient allocation and use of public funds and tenders. It can be accessed from anywhere and having a low lost solution [228] [258]. The specific objectives of the E Procurement MMPs as per Figure 4.9 are as follows:

- Provide faster and cost effective service.
- Improvement in the transparency of public procurement.
- Improve procurement's efficiency.

4.5.5 Electronic Data Interchange (EDI) for Trade (e Trade)

The E TRADE MMP provides following services:

- Enhance the accessibility.
- Transparency improvement in procedures.
- Provide cost-effective service.
- EDI enabled services offered by regulatory and facilitating organizations.
- Introduction of international standards in the field of clearance of export/import of cargo.

The project includes following services [228] [259]:

- Collection and submission of licenses for Foreign Trades.
- Filing and clearance of import export document electronically.

Figure 4.9 Objective of E-Procurement

- Electronic Payment of Foreign Trades license fee.
- E Payment of Custom, ports charges and other airports charges.
- Exchanging the documents electronically among community partners such as Customs, ports, airports etc.

4.5.6 National Portal of India

There are over 5000 websites of Indian administrative departments. Currently citizens have to search any of this website for certain set of tasks. A National Portal of India which contains links of all government authorities could minimize the citizen efforts. This portal acts as a front end to the E Governance program under central/state/UT level [227][228][260]. The objectives of the India Gateway MMP are:

- To ease launch/ application of various E Governance creativities by the Indian Government.
- To emerge as an inclusive one-stop-source of administration information and service transfer through a unified interface.
- To facilitate standard, simple, unified, seamless and general government information access for citizens of India from all treads of life and of various demographic contours.

4.5.7 National Service Delivery Gateway (NSDG)

In National E Governance Plan (NeGP) it is recommended that all governance applications are being implemented in order to deliver speedy services to the citizens at low cost [227]. In order to fulfill this vision, the infrastructure required of NSDG is shown in Figure 4.10.
The objectives of NSDG are [228] [261]:

- Decrease the cost of E Government Projects.
- Achieve interoperability between various department levels of E Governance services which are implemented at different levels.
- Provide transaction logging to track transaction.
- The additional funds requirements for software and hardware could be protected by associating them with alternative platforms.
- Develop Gateway messaging standard and using these standards evolve Central Gateway that is owned by the Government.
- Develop standardization practices under E Governance to enhance interoperability.
- Develop integrated service delivery infrastructure for country-wide operations.

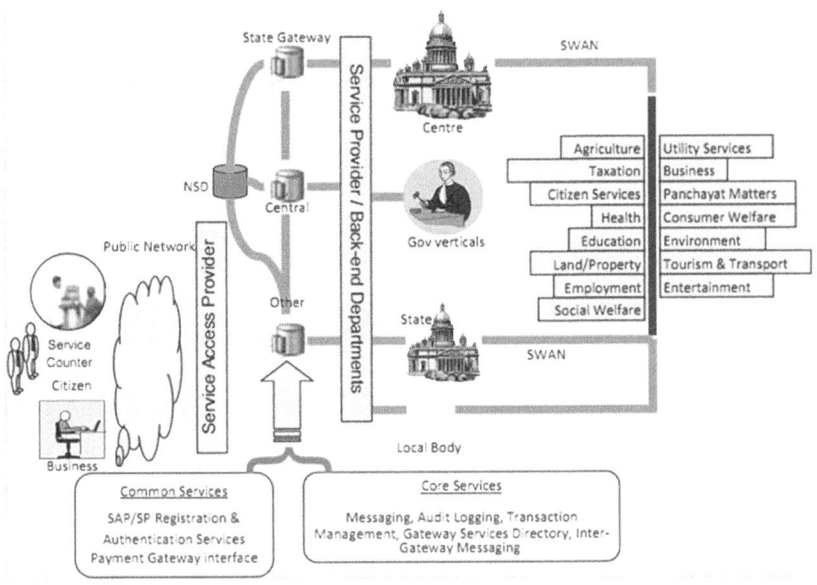

Figure 4.10 Infrastructure of National Service Delivery Gateway (NSDG)

4.6 Conclusion

India has gained a very strong appearance in Information Technology applications globally. The major initiatives are National E-Governance Plan, Common Service Centers, State Data Centers, and Mission Mode Projects. In this chapter, all these major issues have been described to represent a seamless view of Government. The chapter attempts to highlight the features of central, state, integrated categories of mission mode projects which are the future of e-governance in India. The information about all such initiatives will be useful to policy makers, government officials as well as citizens so that we all together can develop competencies and would able to improve the lives of millions.

5

E Governance Data Management Framework

5.1 Introduction

Any E Governance system produces massive data through its various departments. Efficient Data Management policy is required to ensure appropriate data collection, storage and usage. Data could be easily managed by using a single window access so that it could be made available for its all categories of users i.e. administrators, researchers and end users. Various Data Management issues such as data analysis, design, architecture, security, quality and data governance are the major issues highlighted here in this chapter. Data Management framework especially for E Governance is also discussed.

5.2 E Governance Data Management

Data Management is a challenging issue in E Governance and it covers all associated steps of data throughout its life cycle. The need of Data Management was established during starting days of 1980s when high speed data access techniques such as Random Access Disk came into existence. Now, we have terabytes of data for concurrent access and highly responsive real time systems such as E governance where Data Management is essential [262]. E Governance Data Management usually addresses following issues as indicated in figure 5.1:

- E Governance Data Architecture, Data Development and Meta Data Management
- E Governance Data Operation, Security and Quality Management
- E Governance Content Management, Reference and Master Data Management
- E Governance Data Warehousing and Business intelligence.

E Governance Data Center, Data Warehousing and Data Mining: Vision to Realities, 95–112.

Figure 5.1 E Governance Data Management Issues

5.3 E Governance Data Governance

In an E Governance model framework Data Governance deals with all government data, ensure its all time availability and usability in various administrative regulatory and social welfare operations [263]. It also ensures E Governance Data integrity and security to improve the quality of data for better management. The end objective of E Governance Data Governance model is to develop an efficient framework for improved, well maintained and protected information system with well defined set of processes. The Data Governance is also accountable for maintaining trustworthy data for the citizens to facilitate efficiency, transparency and accountability in all government operations through its three role-players i.e. People, Process and Information and Communication Technologies. Following are the two main components of Data Governance:

5.3.1 E Governance Data Asset

Data and information are the two important vital assets of good governance. The success of any E Governance project is mainly based on efficient utilization of data and information [264]. Government organizations utilize tons of data for their decision making. Policy makers are also making use of E Governance data assets to enhance citizen satisfaction and operational efficiency in low cost. Availability of suitable data minimizes the information gap and enhances the operational and strategic activities. At present there is an imperative need of efficient Data Management which could also be done through partnership of business parties.

For seamless exchange of data among systems, two components play crucial role. One is data-exchange and the other is meaning-exchange, which is contextual in nature. Data-exchange is addressed through Technical Interoperability and meaning-exchange is addressed through Semantic Interoperability. Basic elements of Semantic Interoperability are:

- Semantic description of data, which may have contextual meaning.
- Semantic Mediation to resolve conceptual meaning differences, while exchanging the data.
- Semantic discovery of desired assets for seamless exchange of data.

Semantic Interoperability will require an agreement on the precise meaning of exchanged information among the E Governance systems for the delivery of integrated E Services. For this purpose, there is a need for building centralized repository of Semantic Interoperability Assets like:

- XML schema for meta-data of data elements for uniformity in data storage formats of common generic data elements.
- Code lists for controlled values.
- Ontology of common generic data elements in corresponding applications along with its contextual meaning (precise meaning, concept, attributes, constraints, restrictions, business rules, etc.).
- Taxonomies for classification of data.

To ensure uniform mechanism for creation of the repository of the above assets, a prescribed structure should be in place. Further, these assets would be dynamic in nature, due to new additions in the assets and also update of values of standardized data elements from time to time. Hence, there should have to be mechanisms for:

- Version controls of data elements in the repository.
- Maintenance of history of changes in values with time.
- Registration of users of the repository and also process of issuing automatic alerts to the application using repository.

5.3.2 E Governance Data Steward

A Data Steward or Expert is responsible for managing core Data Management strategies and procedures. Figure 5.2 shows the responsibilities of Data Experts. Following are the Data Experts responsibilities:

- Identification of E Governance practices and measures.
- Identification of correctness and consistency of E governance operational data.
- Identification of suitable metadata for E Governance operation.
- Identification of verification and validation process for E Governance data queries.
- Identification of standard documentation and testing practices for E Governance.
- Identification of suitable Data Mart for E Governance.
- Identification of valid user access.
- Identification of best practices for E Governance Data Mining using appropriate Data Warehouse application.

5.3.3 Data Management Challenges in E Governance

The main issue regarding E Governance is to manage the changes such as cultural, operational and processing workflow [265]. There are heavy load of information in global information repository, so main challenge is to focus on achieving interoperability for Data Management. The other issues are:

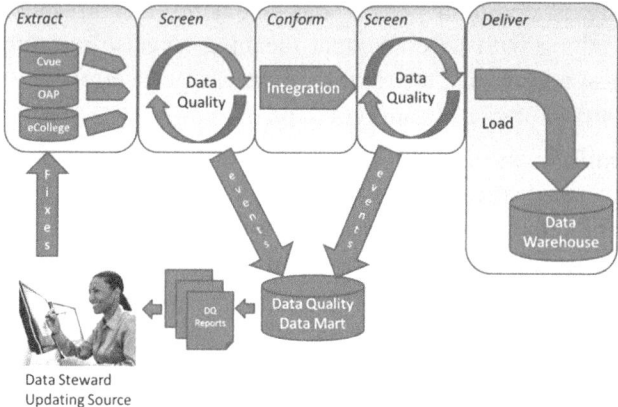

Figure 5.2 Responsibilities of Data Experts

- Common understanding of information is irrelevant due to:
- Heterogeneity in information systems, Database Management System, Operating System, instruction set, coding convention, Digital Media Repository Management System etc.
- Lack of bilateral or multilateral agreements for data exchange.
- Varied legal data systems.
- Heterogeneous formation of legacy systems.
- In communication, for information exchange between the systems, there are usually three participants like:
- System that provides Information.
- System that receives Information.
- System (optional) that monitors information-exchanged between provider (s) and recipient(s), for audit logging or any other reason.
- The conflicts in the contextual meaning are complex, some of which are listed below:
- Data value conflicts.
- Precision conflicts.
- Labeling conflicts.
- Integrity conflicts.
- Conflicts of description of an entity by different attributes.
- Conflicts in mandatory / optional nature of data in applications.
- Ownership conflicts.
- Conflicts in categorization of data etc.

5.4 E Governance Data Analysis

The E Governance Data analysis is a process of establishing quality efforts of E Governance through scrutinizing, changing, modeling E Governance data with the objective of focusing on valuable feedback, suggestions and results [266]. A Common Data analysis technique is Data mining which includes clustering, classification and regression that is based on knowledge discovery in E Governance. Following are types of E Governance data:-

- Qualitative data.
- Quantitative data.
- Categorical data.

In E Governance decision making data aggregation is a key practice. Aggregated data analysis could be done on three different categories:

- Explanatory data analysis: Discovering new features.
- Investigative data analysis: Inspecting new features.
- Authenticated data analysis: Verifying and validation of new features.

5.4.1 E Governance Data Cleaning

In E Governance Data analysis one of the important procedures is data cleaning in which data alteration could be done in order to have correct and error free data. In data cleaning step old data as well new data should be maintained so that roll back is possible at any stage [267]. Data cleaning is a Data Preprocessing technique and essential in E Governance because Data is coming from heterogonous sources. The main objective of Data cleaning is to produce high quality data that should be complete, consistent, valid, correct, uniform and relevant [267]. It could be done by modification, deletion or replacement. The main steps of data cleaning are screening, diagnosis and editing as shown in Figure 5.3.

5.5 E Governance Data Interoperability (EGDI)

Interoperability in E Governance is defined as "the ability of different systems from various stakeholders of E Governance to work together, by communicating, interpreting and exchanging the information in a meaningful way". The interactions between stakeholders should be achieved by sharing of information and knowledge through the business processes supported by them

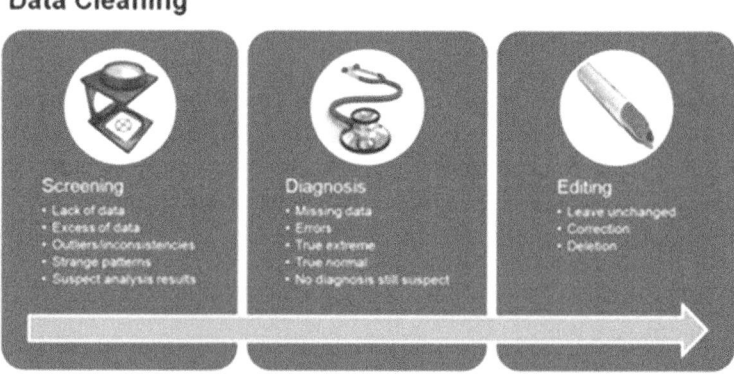

Figure 5.3 E Governance Data Cleaning

[268]. EGDI in India would encompass agreed approach to be adopted by the organizations that wish to work together towards the joint delivery of public services using Information and Communication Technologies, to achieve above mentioned goals, namely exchange of data under set of agreed process. The importance of Data Interoperability for E Governance in India are as follows:

- To provide background on issues and challenges in establishing interoperability and information sharing among E Governance systems.
- To describe an approach to overcome these challenges; the approach specifies a set of commonly agreed concepts to be understood uniformly across all E Governance systems.
- To offer a set of specific recommendations that can be adopted by various stakeholders to pro-actively address the challenges in interoperability.

5.5.1 E Governance Data Interoperability Levels

An EGDI involves a common structure which comprises a set of standards and guidelines; the structure can be used by the public agencies to specify the preferred way that all stakeholders interact with each other to share the information [269]. As shown in Figure 5.4 the interoperability levels related to the sharing of information in EGDI are mainly classified into:

- Organizational Interoperability (like Process-Re-Engineering including Government-Orders, Process Changes, Organizational Structures).
- Semantic Data Interoperability (Enabling data to be interpreted and processed with the same meaning, etc.).
- Technical Interoperability (like technical issues in interconnecting ICT systems and services, information storage and archival, protocols for information exchange and networking, security, etc.); in general, Technical Interoperability was considered for classifying the standards into various layers or domains.

5.6 E Governance Data Architecture

The E Governance Data Architecture illustrates the data structures used with governance applications. It includes data processing, storage and deployment techniques. It also includes standard methods for data flow and their controlling mechanism. The natural life cycle of a data architectural design process is identification of end objectives, continuous monitoring and controlling of

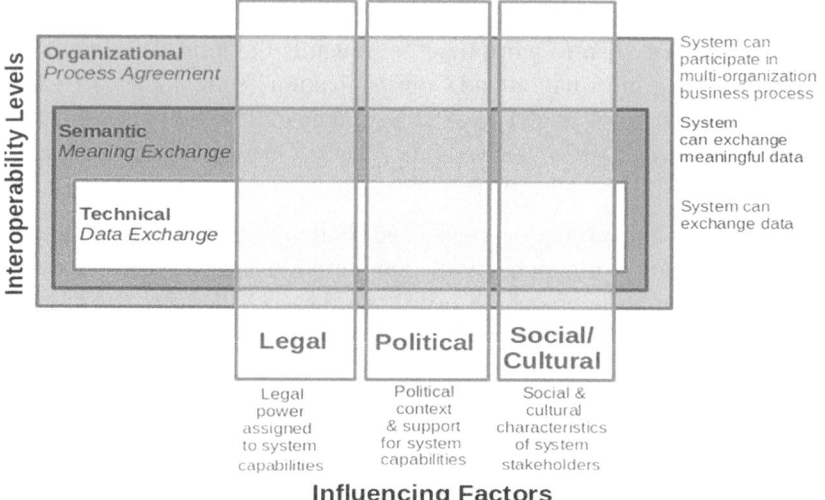

Figure 5.4 E Governance Data Interoperability Levels

the working model and then model enhancement on the basis of feedback. During the identification of end objective of the E Governance Data Architecture the domain has been considered in three different levels such as conceptual, logical and physical level [270]. It covers complete representation of E Governance function, operations, technical knowhow and data assets.

5.7 E Governance Data Modeling

In E Governance, Data modeling is a procedure for exploring E Governance work practices for a database because a data model is ultimately executed in a database. It highlights the data development steps and their utilization in E Governance practices. In a typical E Governance framework the data model would have real time entities with their attributes, data description and associations [271]. The data requirements for E Governance are tagged in the form of associated data definition with their logical data model. Standardization of data modeling is essential in any E Governance application for efficient Data Management, Data Integration and development of Data Repositories. GOI has initiated Public Information Infrastructure with the planned tasks like:

- Encourage the use of Information and Communication Technologies in Public Transport System.
- Proving interconnectivity to all research and educational institutions using National Knowledge Network.
- Employing Information and Communication Technologies in Justice System.
- Promoting Innovation.
- Providing citizen interface in order to promote the public services delivery.

5.8 E Governance Database Management System

An E Governance Database Management System provides compact, efficient storage of government operational data for quicker access with less memory and in more responsive environment. E Governance Database Management System contains logical, descriptive and repeated attributes which needs appropriate processing before decision making [272]. Today the roll of DBMS in E Governance is providing services or features related to data processing, accessing, maintaining and security.

5.8.1 E Governance Data Maintenance

E Governance system consist heterogeneous data in varied applications so there is continuous need of accumulation, cross out, alteration and modernization under data maintenance phase. There are some traditional techniques for maintaining the data either manually or using some tools [273].

Another E Governance Data Maintenance issues are handling process changes, managing staff and users trainings, enhancing wider participation of stakeholder for all emerging technologies including various upcoming features of E governance.

5.8.2 E Governance Database Administration

E Governance Database administrator is important for managing the data and it is responsible for database maintenance and working [274]. Administrator having following qualities:

- Testing support.
- Reliability.
- Security.
- Availability.

- Recoverability and backup plans.
- Performances measure and taking high quality decisions.

5.8.3 E Governance Data Access

E Governance data access is a set of data related operation such as saving, fetching and querying with data repositories. Since Data repositories includes different file types so different accessing methods such as sequential access and random access is used for accessing the data [275]. Currently in various E Governance applications standardized languages such as SQL, ODBC, JDBC and XML has been used for E Governance Data access services. The figure 5.5 is showing the complexity involved in data access mechanism in E Governance.

5.8.4 E Governance Data Erasure

Data erasure is a permanent data deletion method that totally destroys all electronic data. Data is one of the crucial assets of any organization therefore it is important to protect the data from unauthorized access and attack [276].

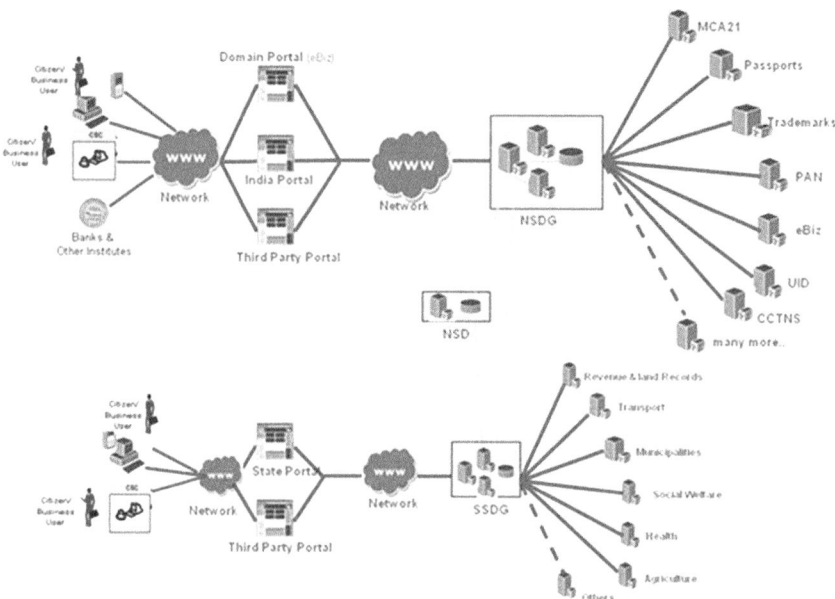

Figure 5.5 E Governance Data Access

5.8.5 E Governance Data Privacy

Data privacy is a strongly recommended within any organization. Data Privacy in E Governance is concerning issue among citizens, politicians and legal practitioners. In E Governance maintaining Data Privacy is a challenging task because all government system shares various departmental data and also protects some department level information [277]. E Governance Data Privacy is crucial in following departments:

- Hospitals and Healthcare centers.
- Judiciary systems.
- Police and criminal investigating systems.
- Banks and financial organization.

5.8.6 E Governance Data Security

Data privacy could be achieved through Data security. It prevents unauthorized data access and ensures permanent Data storage [278]. Common Data security techniques are Data Encryption and Decryption.

5.8.7 E Governance Data Identification

Various government agencies will recognize the data, define and maintain their metadata into data catalogue for future identification purpose [229].

5.8.8 E Governance Data Authenticity and Retrievability

Government agencies will supervise the data and business documents to preserve data authenticity and retrieval to satisfy business and statutory requirements [278].

5.8.9 E Governance Data Auditability

Data elements should be defined in proper and concise manner and stored in consistent format and content must be recorded in an audit Trail [279].

5.8.10 E Governance Data Replication and Interfaces

The data is replicated among various agencies within the governmental organizations along with efficient interface [280].

5.9 Data Quality Management

The objectives of E governance are quicker citizen services, efficient decision making and planning. All these could be achieved by high quality data. Alternatively, the data are deemed of prime quality if they properly represent the real-world construct to that they refer [281]. In addition to this, if data volume increases, data consistency is required to be maintained. The various characteristics of data quality are highlighted in Figure 5.6.

5.9.1 E Governance Data Integrity

In an E Governance task it is necessary to possess complete and consistent data for any government operation in such a way that it might able to reflect complete information and guaranteed integrity of data in use [282]. Data with appropriate integrity is required to maintain throughout the operations (such as transfer, storage or retrieval). Data integrity also assures data consistency, which could be certified and reconciled.

5.9.2 E Governance Data Quality Assurance

Data quality assurance consist a set of actions to find out data inconsistencies and other anomalies to achieve high quality [281].

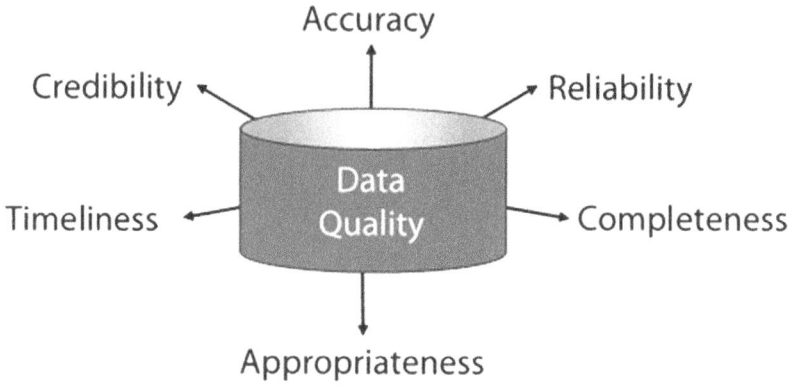

Characteristics of Data Quality

Figure 5.6 E Governance Data Quality

- Ensuring consistency so that information exchanged is interpreted and processed unambiguously in all the interacting-systems at all the time.
- Building trust that the communicated information from multiple sources is valid, and without any ambiguity, especially in automated systems. In case of human communication it tolerates considerable ambiguity for applying wisdom to have meaningful information.
- Ensuring reproducibility and reliability of the data that is collected or encoded at various sources and reproduced for the usage among the stakeholders. This holds both for individual and aggregated data at organization level.

5.9.3 E Governance Data Integration

In Data integration data collected from different sources may merge to provide a unified view [282]. Data integration is commonly required in different commercial and scientific applications.

Following are the layers of Data Integration as indicated in Figure 5.7:

1. Presentation: This layer provides an interface for storing and retrieving data and accessing information to the user. This concerns most of the user interface aspects related to presentation of information in various formats. This includes standards and technologies that present data to the user in various means of access (personal computers, smart cards, mobile phones, PDA hand-held devices, digital televisions, etc.) especially for the purpose of E Services. This

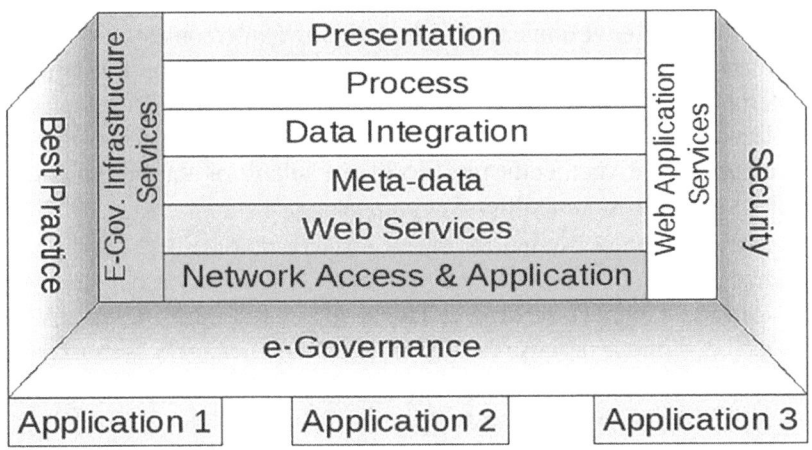

Figure 5.7 Layers of Data Integration

layer is divided further on the basis of service delivery mode and standards. The common standards found in this domain include: HTML, XHTML, WML, etc.

2. Process: This layer provides standards of various software and business processes. Government Business processes is needed to be aligned for proper inter-operation and integration. For example, a process describing how to file income tax return in batch processing using web services provided by income tax department website.

3. Data Integration: This layer integrates the data from different systems such as heterogeneous and homogeneous system to provide consolidated view of the data. Data integrity and consistency is to be maintained. Data Integration layer have various standards and technologies for storage, accessing and management government information. The common standards found in this domain include XML and UML.

4. Meta Data: This domain deals with the core standards required for describing the data structures and their mapping to real-world entities, relational database table structures, XML schema used in the systems. The common standard(s) found in this domain include: Dublin Core Meta-data Initiative (DCMI).

5. Web Services: This is the fifth layer of Data Integration. It provides standards that support for exchanging the data and services across homogeneous and heterogeneous systems.

6. Network, Access and Application: Network layer consist interoperability components which support the information exchange and communication across distributed environment. It determines the interconnectivity of various information processing resources. Technical specification for achieving the interoperability among various applications is supported by the access layer. The application layer consist standards for recognizing communication partners, determine whether the resources are available or not and synchronize the communication among different components.

7. Security: This layer support security services that enable security when different components communicate with each other and various domain of E Governance Architecture Model. This deals with security issues like encryption, decryption, passwords, digital signatures, etc. The security layer controls all technical interoperability layers. It includes various standards and technologies which provide secure exchange of information, provide secure access to public sector information and also protect the identity information of users and resources.

5.10 E Governance Master Data Management

Master Data Management (MDM) contains a detailed action plan for gathering, aggregating, matching, joining, quality-assuring, storing and sharing of data among various organizations [272]. The aim of Master Data Management is to ensure consistency and control in various data related operations.

The structure for the Management of EGDI used by the public agencies in the global scenario for providing public services has been established in any combination of the following ways:

- Existing approved committee.
- Developed a new committee that supervises the interoperability scheme.

Management structures are developed in a number of ways for example using state law or by joint agreements that is signed by the agencies of different jurisdictions.

The E Governance initiatives are aligned with:

- E Governance Strategy of Government.
- Legal requirements (data protection and privacy information of the citizen, etc.).
- Administration and custodianship of public agencies with reference to information management.
- Executive orders and laws related to E Governance services.
- Citizen services driven Administrative Procedures Enforcement, etc.

The interests of both government and individual public agencies are captured and balanced through the Management of EGDI. The Government of India (GOI) created the 'Apex Body' under Department of Electronics and Information Technology (Deity) to play a critical role in the formulation and ratification of E Governance standards and related guidelines and procedures. Apex Body's main functions include preparing road map with time lines for E Governance service deliveries, priorities, goals, approve budgets etc.

5.11 Intelligence Decision Making Capability in E Governance

Intelligence decision making applications are a set of technologies for assembling, storing, investigating and accessing data for decision making [283]. In general such applications include, query processing, On Line Analytical Process (OLAP), Prediction, Decision Support Systems and Data Mining.

Intelligence decision making consequently used in E Governance applications including various government to government and citizen related operations. A Data Warehouse is required to provide an efficient repository which could provide necessary foundation for government decision making process.

5.12   E Governance Document, Metadata and Identity Management

5.12.1 E Governance Document Management System

E Governance Document Management System (EGDMS) is an automated process used to track and store various government documents [284]. It also keeps track of the various versions produced by various users. It has an Electronic records management systems which tracks and store records with security and auditing features for example office documents, e-mails, certificates such as birth certificates, Medical records etc.

5.12.2 E Governance Meta-data Management

Meta-Data Management is a repository management system which keeps data about data. It has two basic steps i.e. Metadata Discovery and Metadata Publishing. Metadata discovery is the method of discovering the semantics of a data element in data sets and performs mappings between the data source elements and a central metadata registry [285]. In Metadata publishing the availability of data elements could be ensured to external users.

5.12.3 E Governance Identity Management

Identity management in E governance is a way to identify various entities in E Governance system (such as a country, a state, or an organization) and managing controlled access to the resources [286]. Identity management includes many dimensions, such as:

- Technical.
- Legal.
- Social and humanity.
- Security.
- Organizations.

5.13 Conclusion

E Governance resolution is the source of huge data generation through government agencies. The data are used in various departments in their own formats so it is necessary to achieve the interoperability among them and provide a management framework that not only manages the huge data but also provide secure access, privacy and maintenance [287]. Data Management is also important because a central database needs commonly accepted data format after appropriate data cleaning and transformation. All theses phases must be handled efficiently under universal policy framework so that a central database could be developed for their Data Mining operation. The key objective of Data Management is to remove data inconsistencies and ambiguities for correct Data Warehousing and Data Mining operation.

6

E Governance Data Center

6.1 Introduction

Latest developments in Information and Communication Technologies and emergence of E Governance arises the need of investment in technology resources to achieve the promises of good governance. This also needs holistic management capabilities and efferent interoperability between IT tools and government organizations. Data center facilitates the E Governance system with huge repository which consists millions of megabytes of data [288]. The huge data repository could be accessed by multiple departments for their day to day operations. The important components of Data Center are group of servers and networking devices. The group of servers is for storing the departmental data and Networking Infrastructures is for making them accessible to all authorized stakeholder. Data Center is an upcoming trend within E Governance system is growing tremendously but also facing challenge in terms of reliability, cost and security. In this chapter need of data center in E Governance system has been established. It is also highlighted that cloud computing is an emerging technology and also utilized by the data center for different services. Various infrastructure requirements of data center are also discussed in this chapter along with their associated challenges. Design, architecture, security, quality and data governance are the major issues highlighted here in this chapter. Data Management framework especially for E Governance is also discussed.

6.2 Use of Data Center in E Governance System

The core function of E Governance is to provide government services round the clock through single window system. The necessary information could be provided quickly only if there are adequate amount of reliable data. The performance and reliability of E Governance system depends upon the standardization of data center [288]. It is also very difficult to collect data from

E Governance Data Center, Data Warehousing and Data Mining: Vision to Realities, 113–128.

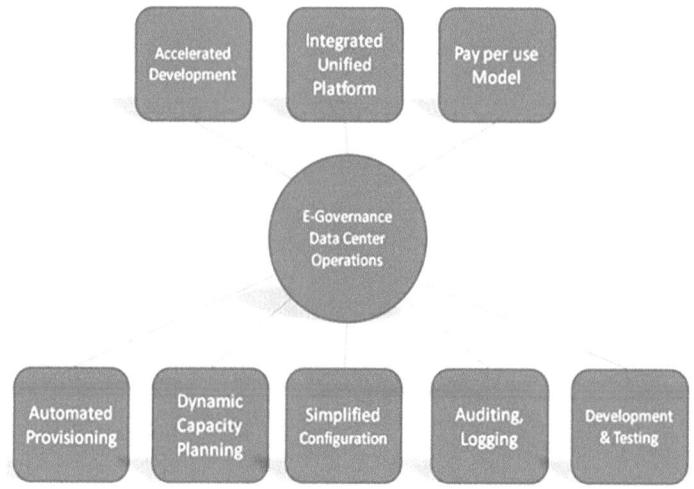

Figure 6.1 E Governance Data Center Operation

heterogeneous environment and then converting the data into common format so that it could be shared by all the government departments. The data collection and data sharing need proper authentication and authorization under a secure environment so that the security, availability and performance could be maintained.

The E Governance Data Center is a centralized, consolidated and secure repository for all E governance operational data. The Data Center could be observed from the viewpoints of all its stakeholders such as Citizens, Government Officials, Administrators and Policy Makers as highly functional, reliable, interactive and an efficient platform for their seamless coordination [288]. The basic infrastructure of Data Center typically comprises collections of multiple servers and network switches and various other application programs to serve the purpose of E Governance System. The Figure 6.1 shows the details of E Governance Data Center operation.

In any E Governance project Data Center is essential due to following reasons [288] [289]:

- It facilitates the feature of Online Government so various operations could be performed easily using Single Window Interface.
- Data Center serves the purpose of shared repository so various departments can share the valuable data and information through Data Center.

- Data Center could be developed with strong backup support so high availability of government services could be ensured on the basis of 24x7.
- In any E Governance services security is primarily required which could be achieved by using Data Centers.
- A secure, transparent and efficient common E Governance service platform could be provided through Data Centers.
- Data Center helps us to offer well-organized electronic delivery of services between Government, Business and Citizen.

6.3 E Governance Data Center Implementation Using Cloud Computing

Cloud Computing offers enormous opportunities for E Governance applications globally, to provide efficient, reliable, fast governance with lower overheads. E Governance services could be offered more powerfully with the help of Cloud features such as application virtualization, service management and instant deployment [288][290]. With the assistance of correct designing, execution, training and sensible management, the Cloud infrastructure would greatly cut back overall prices of various departments for maintaining and managing their services under E-Governance [290][291]. The Figure 6.2 demonstrates the uses of cloud over government domain.

There are essentially four categories of cloud computing services which support the establishment data center for e-governance [288][292]:

- IaaS: Infrastructure as a Service
- PaaS: Platform as a Service
- SaaS: Software as a Service
- MWaas: Middleware as a service

6.3.1 IaaS, Infrastructure as a Service

E Governance services needs round the clock availability of infrastructure without having any interruption. There are millions of users interacting with the government on day to day basis. This needs powerful infrastructure such as unlimited processing capabilities, storage and bandwidth. The cloud computing Infrastructure as a Service manages all these features of E Governance with scalability [288]. For example the Indian railway portal requires a server with 24 hours uptime and online services to users. This could be implemented

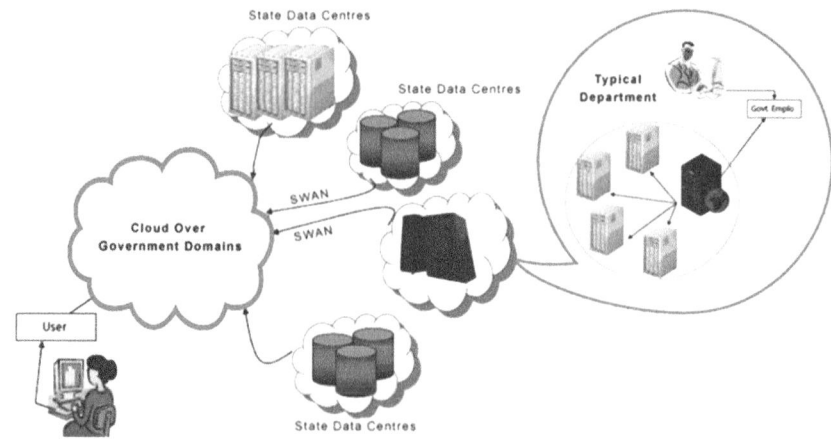

Cloud Scenario for e Governance Application

Figure 6.2 Cloud Scenario for E-Governance Application

by cloud computing using IaaS model so that unlimited processing power, storage space and bandwidth could be obtained [290][292].

6.3.2 PaaS, Platform as a Service

In E governance automation of any department may need deployment of new operating system and database management system [290]. The cloud computing with Platform as a Service could easily support these entire requirement with minimum overheads [292]. Following are the features of the PaaS:

- On Demand installation of Operating System.
- Customized Query Processing System.
- Database Software services as per E Governance project.
- On demand Workflow Services.

6.3.3 SaaS, Software as a Service

Cloud computing with software as a service considers various E Governance applications as a service [288]. Suppose a citizen centric service is required to be implemented in E Governance , all hardware and software related solution could be efficiently obtained by the SaaS feature of cloud computing[290][292]. Following are the services that could be easily implemented through Cloud computing:

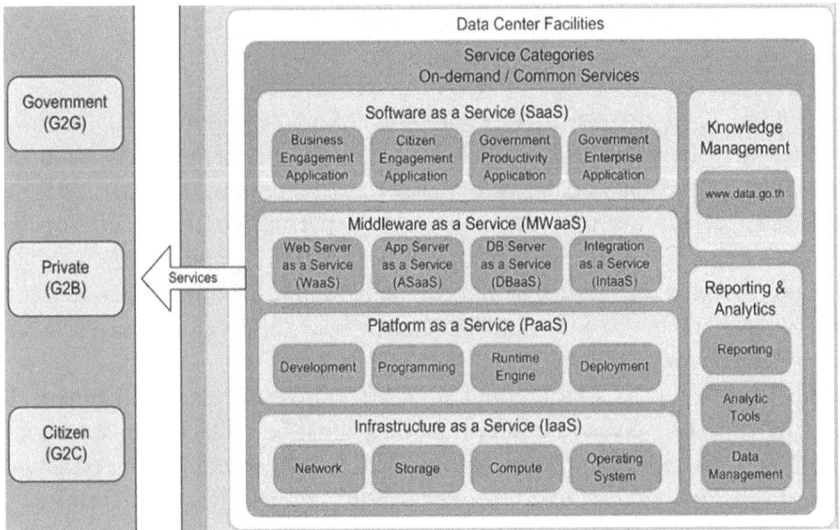

Figure 6.3 Data Center Services through Cloud Computing

- Grievance Readdressal System.
- Pay Roll Management Systems.
- Human Resource Management Systems.
- E-police, E-court.
- Rural/Urban Municipalities Automation System.
- Payment Gateways for Water, Electricity and basic amenities.
- District/Block Management System.
- Single Window Citizen Interface.

MWaas, Middleware as a Service: Middleware is special purpose software that offers some additional software features which is not available in a typical operating system. It provides a simple interface for communication and input output so that the software developer could concentrate in the development of core software functionalities. Middleware as a Service offers cheaper and quicker development and deployment environment with all runtime requirements with minimum complexities of the essential hardware and software components [293][294].The Figure 6.3 illustrates different layers if cloud computing within Data Center especially deployed for an E Governance applications.

6.4 E Governance State Data Center Using Cloud Computing

The Figure 6.4 explains cloud computing implementation for data center services. Users in a ministry put their request for services then the request goes to the cloud data center through internet or private network [291]. In cloud data center the user authentication is performed for permitting the required action [288] [295]. A mapping system and an application manager is responsible for implementing the core task required by the user and access control mechanism is responsible for sending back the service to the requester with proper security[291].

The data center is a vital component of E Governance establishment. In any E Governance System appropriate data must be collected in secure environment [296]. The proposed data center using cloud computing paradigm is aimed to facilitate availability, accessibility, reliability and security of E Governance data with compete power back and high disaster recovery in cost effective manner. There are still some more challenges in cloud computing like integrity and privacy [288].

6.5 Three Tier Architecture of E-Governance Data Center

Data Center is a secured central repository which is suitable for government data storage for future utilization. It consist powerful storage servers which store the data from various Departments/ Ministries in the government and efficient networking infrastructure for reliable communication and data transfer among various departments/organizations [288]. As shown in Figure 6.5, E Governance Data Center consists three main parts:

- Users
- Middleware
- Servers

The user could be a citizen, government officers or policy makers. In E Governance system each ministry operations are handling data that is sensitive and important. These ministries also share data as per their agreed upon terms and policies so the sharing of data could be managed efficiently through Data Center. Data center are useful because it provides data protection from likely threats or attacks and also manage interoperability among the different set of data [288].

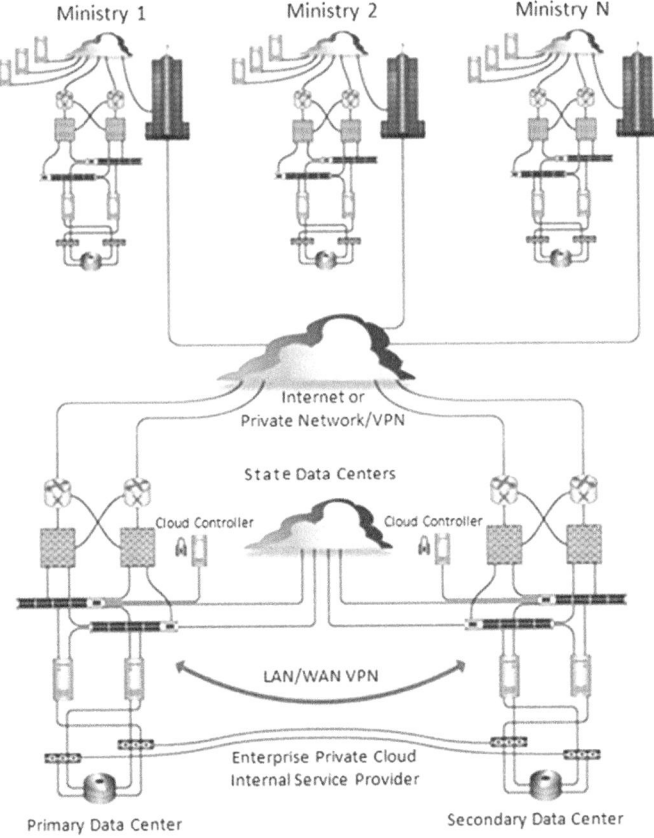

Figure 6.4 E-Governance Corporate Data Centers

Middleware is also another important component of Data Center. It consist Service Level Agreement and Access Control Management [288]. Each ministry of government avails the services of data center duly supported by Access Control Management. Access Control Management manages various users along with their access profiles and rights. It also keeps different data format with their appropriate access techniques. It also manages the whole data center services which are offered to various government ministries.

Data Center consist multiple servers which stores different types of data [288]. The Servers need to share their data and information to get some job done. In E Governance Data Center servers are categorized in terms of their usage as well as also in terms of different department/ Ministries.

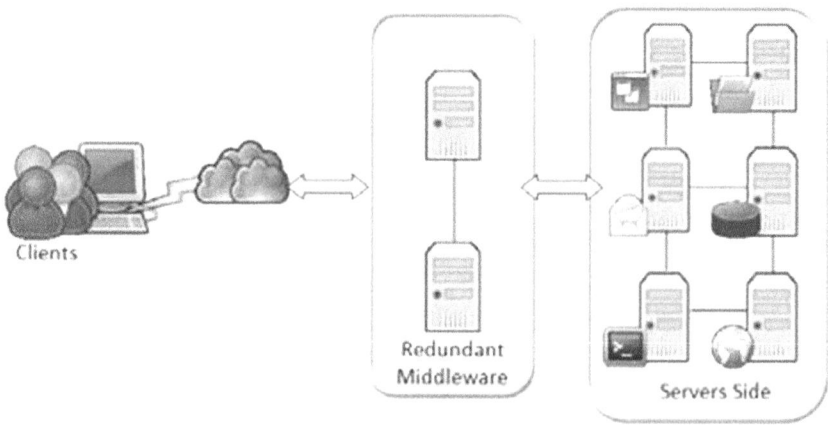

Figure 6.5 Data Center Layers

ZONE A	ZONE B	ZONE C
Server racks, Networking racks, Structured cabling racks, Storage Area Network box, High end Servers, etc	• NOC Room (Network Operation Centre) • Centralized Building Management Systems (BMS) monitoring room • Help-Desk Area • Testing / Lab Room	• Electrical Room (Power Supply room) • Telecom Room • UPS and battery room • AHU • Fire Suppression System

6.6 Infrastructure Details of Data Center

The Data Center needs adequate office space in a centrally located area of an organization [297]. The data center infrastructure can be primarily divided into 3 basic divisions [298]:

- Physical infrastructure
- Application infrastructure
- IT components

6.7 Physical Infrastructure of Data Center

There are three categories of data center on the basis of infrastructure required [291] as mentioned below:

6.8 Detailed Infrastructure of Data Center with Respect to Zonal Divion

Air conditioning: A dedicated air conditioning system is essential for temperature regulation in Zone A and a shared cooling system could be used for Zone B and C [298].

Quality concern of all Equipment: In all categories of Data Center, ISO 9001 quality assurance standard is used.

Air Filtration: Air filtration in Data Center is performed by using pleated cell filters.

Electric Re-heater: It is used to prevent failure against overheating.

Humidifier: It is used to control humidity level in Data Center for proper functioning.

Power Panel: This unit is responsible for uniform power distribution among various units of Data Center [297] [298].

Microprocessor controller Panel: This unit monitors the working of Data Center and generates alarm in case of emergency [298].

Availability electricity distribution system: It provides dual power supply to avoid downtime during maintenance.

Switchboard: Switchboards and panel board are essential to support different needs of equipments in terms of load [297].

Lighting: Enough illumination is required in a Data Center for smooth functioning.

UPS System: It ensures round the clock availability of power backup at all loads [297].

Generator: A powerful generator is required to maintain the power back up.

Surge Protection System: Surge protection is essentially required in a Data Center.

AMF Panel: It provides an interface between generator and UPS during power failure.

Video Surveillance: Round the clock close circuit cameras are required for security of Data Center.

Access Control: Efferent access control is required to prevent unauthorized access of Data Center.

Raised floor and insulation: It is required for rigid and stable setup of Data Center.

Fire Detection: It is essential to have smoke detections system along with fire alarm panel.

Fire Suppression: It takes care of Data Center area under critical and non Critical categories.

Pest Control System: It is also necessary to have clean work place for Data Center without having any Pest.

Architectural Work: All building requirements must be fulfilled to have complete set of Data Center office.

Monitoring System: A centralized monitoring system is used for Data Center equipments.

6.9 Application Infrastructure

E Governance data center is the key requirement for the automation of all projects of Government Systems, Departmental Websites, Citizen Portals, Kiosk supported by huge databases. All these features are to be installed into various application servers [298]. Following are the details of application servers:

Web Servers: Web based portals and websites are the easiest way of making the information available for all. All websites needs web servers for hosting purpose. A Data Center with web server is helpful for hosting any web based applications which could be accessed from the internet or any government intranet and extranet.

Application Server: Automation of different departments needs application software so that necessary work flow could be maintained. An appropriate middleware with required application server provides an efficient interface to the end user. The integration of web server and application server is essential to provide high accessibility and performance so that more number of users could be served with suitable load balancing.

Database Server: The database server is required to store and retrieve E Governance operational data so that citizen services could be ensured without having any delay. The Database server is needed to be integrated with web server and application server for various online G2C services.

Directory Server: Directory servers consists all verification and validation policies for users. This would also responsible for security management issues especially at the time of multiuser environment and improve protection, reliability and overall efficiency.

Proxy Server: The Proxy servers are responsible for highly accessible internet facilities with appropriate bandwidth and web-caching. With the help of this server fast and secure online transaction could be performed.

DNS/DHCP Servers: DNS server consist the directory of all domain names and their respective IP addresses which is hosted for public access. The DHCP server performs dynamic allotment of IP addresses to various devices /hosts. It also authenticates users during remote access.

Management Server: The management server manages all distributed systems in the network and ensures efficient and reliable usages of inventory and other computing resources.

Intrusion prevention system: This would provide proactive information over a network to detect and log into a database.

Staging Server: It is an additional server used for some primary task and installed just before the production server.

Server load balancer: It is used for load balancing within the group of clients as well as server machines.

Backup Server: All data must be replicated by using a backup server with the help of an automated process.

6.10 IT Components

Data centers contain a set of routers and switches for data traffic among them and outside world. The IT components details are given below [298]:

LAN Switch: It is used to create Local Area Network among various machines serving for Data Center establishment.

Access switch: It is used to provide an intelligent access of network and devices for authorized users.

Internet Router: It identifies the optimum path for traffic management in internet.

Network Intrusion Prevention System: It prevents the system from any illegal access or activities across the network.

Host based Intrusion Prevention System: It prevents the system from any illegal access or activities within the system.

Perimeter Firewall: This enables the Prevention System to prevent hacking attacks from the internal network.

Internal Firewall: This enables the Prevention System to prevent hacking attacks from the outer world of internet.

EM64T Server: The server, workstation and desktop platforms are combined with supporting software.

Blade server: It minimizes the space and energy requirements of the system.

Blade Specifications: It is an auxiliary unit required to install the servers.

Database Server: It provides all database related services for the system.

SAN Switch: It creates and maintains the channel required for communication.

SAN Storage management software: It helps to maintain the storage systems of Data Center.

Tape Library: It contains tape drives for storage purpose.

Backup software: It is used to perform backup of Data Center.

Management and Monitoring System: It works as administrator for management and monitoring of the Data Center.

Workstations: It is a powerful computer system used for some specific purpose.

Directory Services: It create, store and maintain directories in Data Center.

Antivirus: It is used to provide a safe guard against viruses.

Syslog server: It Provides computer data logging facilities for message generation.

Proxy server: It is an additional server responsible for handling users queries.

KVM: It consists Keyboard, Mouse and display devices for input output purpose.

Data Cabling: It is a physical media provides connectivity.

UTP Cable: It is a type of cable which is free from electromagnetic radiation.

Fiber Optic Cabling Systems: It is a type of connecting media made up of optical fiber

Fiber Optic Connectors: It is used to join two fiber optic cables.

Fiber Optic Patch panels: It is used to terminate a fiber optic cable.

6.11 Data Center in India

Data center has become most valuable asset of any organization i.e. private, public, government. It is an upcoming technical advancement and still updating on day to day basis as per organizational need. In India the current status of data centers have been indicated by using Figure 6.6 [299] [300].

State Data Centers are multiple data centers established in various states of India to provide fundamental IT infrastructure for various E Governance programs [300]. The primary purpose of these data centers is to provide a basic physical facility for various state level E Government applications. The SDC project approved is a part of the National E Governance Plan.

State data center and its maintenance profiles [300]:

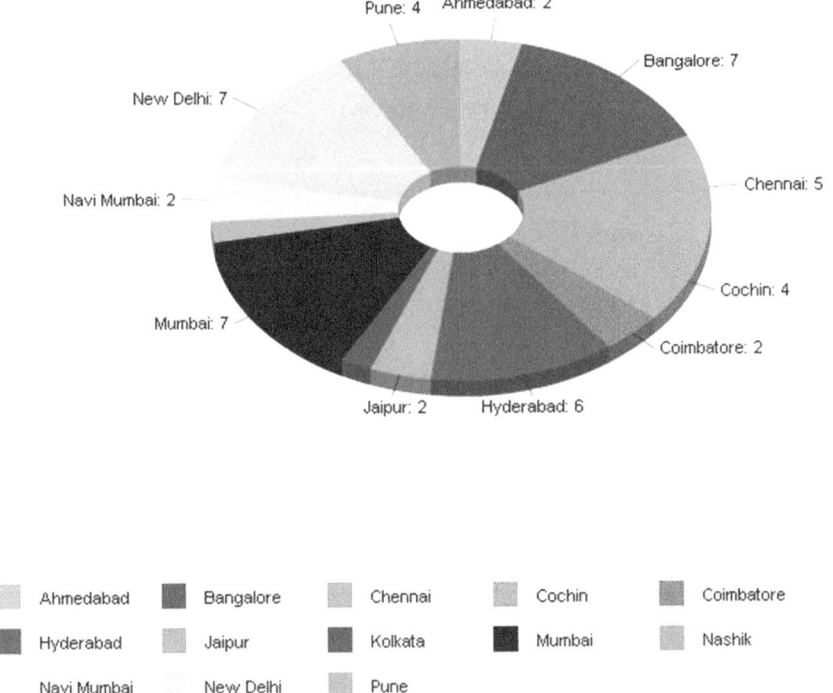

Figure 6.6 Data Center Details in India

6.12 Challenges in Cloud Computing for E Governance

Cost is the prime challenges in e-government system especially for data center [301]. The world is facing an economic crisis that costs big factor. To handle such challenges, three categories of cost for data center are defined as follows [288]:

- **Development cost:** It is the cost that requires making master plan, building infrastructure, buying hardware and software, making database, building security mechanism etc.
- **Operating cost:** It is the cost of operating the system. It includes cost for electricity, cooling system, hiring manpower all the time, paying software license, network cost etc.
- **Maintenance cost:** It is the cost of maintaining the system. Since system life time is large and cyclic in nature in E Governance so that the software has to be modified, updated and changed frequently and similarly

S. No.	State/UT	State Implementing Agency (Nodal Government Agency)	State Consultant	Data Centre Operator (DCO)	Third Party Auditor (TPA)
1.	Andaman	SOVTECH (Society for Vocational Technical Education)	PWC	TCS	E&Y
2.	Andhra Pradesh	Andhra Pradesh Technology Services Limited (APTSL), Hyderabad	PWC	Wipro	KPMG
3.	Arunachal Pradesh	Arunachal Pradesh State Council for Science and Technology	PWC	-	KPMG
4.	Assam	Assam Electronics Development Corporation Ltd (AMTRON), Assam	PWC	-	-
5.	Bihar	Bihar State Electronics Development Corporations Ltd (BSEDC)	3i- Infotech	TCS	-
6.	Chattisgarh	Chattisgarh Infotech Promotion Society (ChiPS), Chattisgarh	3i- Infotech	Sify	-
7.	Goa	InfoTech Corporation of Goa Ltd, Goa	PWC	-	-
8.	Gujarat	Gujarat Informatics Limited, Gujarat	PWC	Wipro	E&Y
9.	Haryana	Haryana State Electronics Development Corporation Ltd (HARTRON)	PWC	Prithvi	Deloitte
10.	Himachal Pradesh	Society for IT and e-governance, HP	3i- Infotech	-	-
11.	J and K	Jammu and Kashmir E GovernanceAgency (JaKeGA)	Wipro	Trimax	-
12.	Jharkhand	Jharkhand Agency for Promotional Information (JAPIT), Jharkhand	PWC	Sify	-
13.	Karnataka	Centre for E Governance(CEG)	PWC	TCS	KPMG
14.	Kerala	Kerala State Information Technology Mission, Kerala	Wipro	Sify	E&Y
15.	Lakshadweep	Lakshadweep Information	PWC	-	-

		Technology Services Ltd (LITSS)			
16.	Maharashtra	SETU Maharashtra (State Level Apex SETU Society of Government of Maharashtra)	PWC	Wipro	Deloitte
17.	Madhya Pradesh	Madhya Pradesh State Electronics Development Corporations (MPSEDC), Madhya Pradesh	PWC	HCL	-
18.	Manipur	Manipur Science and Technology Council, Manipur	Wipro	Reliance	-
19.	Meghalaya	Meghalaya Information Technology Society (MITS)	PWC	Sify	PWC
20.	Mizoram	Mizoram State E GovernanceSociety	Wipro	Prithvi	PWC
21.	Nagaland	Nagaland e governance society (NeGS)	3I-Infotech	Prithvi	-
22.	Orissa	Orissa Computer Application Centre (OCAC), Orissa	Wipro	Spanco	E&Y
23.	Puducherry	Puducherry E Governancesociety (PsGS), Puducherry	PWC	TCS	Deloitte
24.	Punjab	Punjab State E GovernanceSociety, Punjab	Wipro	-	-
25.	Rajasthan	Rajasthan State Agency for Computer Services (RajCOMP), Rajasthan	Wipro	Spanco	KPMG
26.	Sikkim	Centre for Research and Training in Informatics, (CRTI), Sikkim	3I-Infotech	Sify	PWC
27.	Tamil Nadu	Electronics Corporation of Tamil Nadu Ltd (ELCOT), Tamil Nadu	PWC	Wipro	E&Y
28.	Tripura	Tripura State Computerization Agency, Tripura	3I-Infotech	Sify	PWC
29.	Uttar Pradesh	Uttar Pradesh Development Systems Corporation Ltd, Uttar Pradesh	Wipro	TCS	Deloitte
30.	Uttarakhand	Technology Development Agency, Department of Information	Wipro	-	-
		Technology, Govt of Uttarakhand			
31.	West Bengal	West Bengal Industry Development Corporation Ltd (WBEIDC)	PWC	Wipro	KPMG

hardware has to be replaced, configuration has to be changed. The cost that is required in these activities belongs to maintenance cost. It is well known that maintenance cost is more than the development cost.

In order to provide the backup system we need to have a separate redundant data center working in parallel which is a costly affair [288] [301]. It is also expensive to keep data center safe from natural disasters like typhoon, earthquake and tsunami. This issue has given extra stress to the people who are working in the position of system engineer and system design in terms of providing an efficient disaster recovery mechanism for data center.

6.13 Conclusion

In E Governance, the potential application of Data Mining and Data Warehousing could be utilized by well-organized data management through powerful Data Center. Cloud computing is an emerging area providing solutions for Data Center establishment, functioning and maintenance. Here, in this chapter the importance of cloud computing in E Governance Data Center has been discussed. The various requirements of data Center deployment such as physical, application and network infrastructure is also described extensively. Although cloud computing is a revolutionary technique especially for large organizational environment such as E Governance, it is facing challenges. This needs highly skilled man power, substantial one time establishment cost and long term recurring operational and maintenance cost. A low cost open source Cloud Computing solution may be adapted for typical needs of E Governance system.

7

Data Warehousing and its Applications in E Governance

7.1 Introduction

This chapter highlights the basics of Data Warehousing and its application in E Governance with the help of a novel E Governance model framework. The chapter indicates the need of Data Warehouse in E Governance system with its possible benefits and drawbacks. It also proposed an E Governance model framework where the first part represents Administration Module and the second part represents Technical Knowhow Module which proposes framework of Data Warehouse and State Data Center. The third part includes Service Block and the fourth describes Stakeholder Block of the proposed model. The concluding observations are given at the end.

7.2 Fundamental of Data Warehousing

A Data Warehouse is a database repository used for exploring the data. It serves like a central repository that collects data from different sources. Generally Data Warehouse contains historical data that is generated by operational systems. It helps the professionals to improve their decision making capabilities through queries and analysis. For example, a Data Warehouse contains large amount of information regarding customers' behavior, their requirements to analyzing the business requirements for quick decision making. A classical Data Warehouse consists following components as shown in the Figure 7.1

- **Operational System:** It consists Single/Multiple heterogeneous data source(s) such as Banking System, Enterprise Resource Planning System and Government System etc.
- **Integration Layer:** The integration layers collects and integrates the data from different sources and then converts it into operational data

E Governance Data Center, Data Warehousing and Data Mining: Vision to Realities, 129–154.

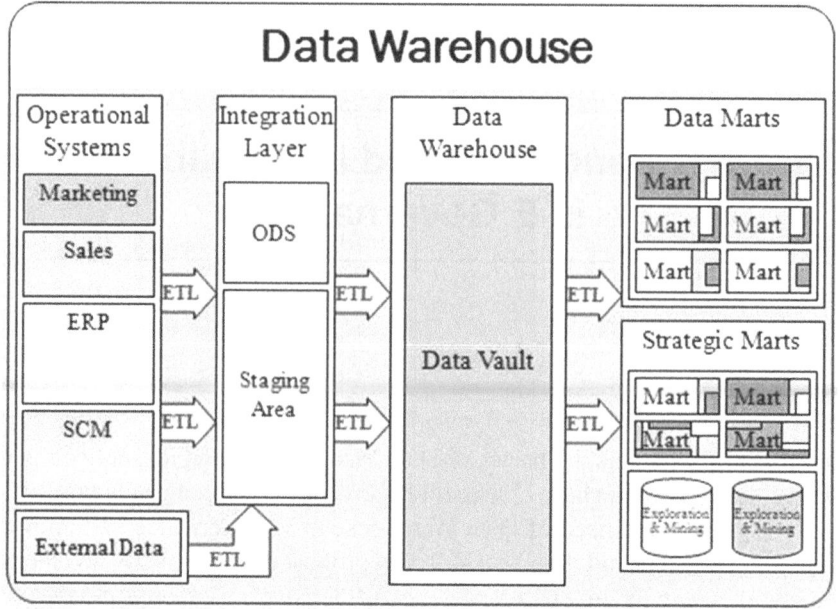

Figure 7.1 Fundamental Data Warehousing Diagram

store (ODS) and transformed to Data Warehouse. The staging layer also collects unprocessed data that is fetched from multiple data sources.

- **Data Warehouse:** It is a central repository which is organized into hierarchical groups according to their facts and dimension
- **Data Mart:** A logical and simplified subset of Data Warehouse used for Data Mining.
- **Extraction, Transformation, and Loading (ETL):** It performs cleaning and transforming of data.

7.3 Data Warehouse Architecture

The technical architecture of a Data Warehouse is similar to other data systems. The architecture for Data Warehouse designed for real data scenario is strictly depending upon the specific situations of any organization [302]. There are mainly 3 basic architectures in data warehousing, mentioned as follows:

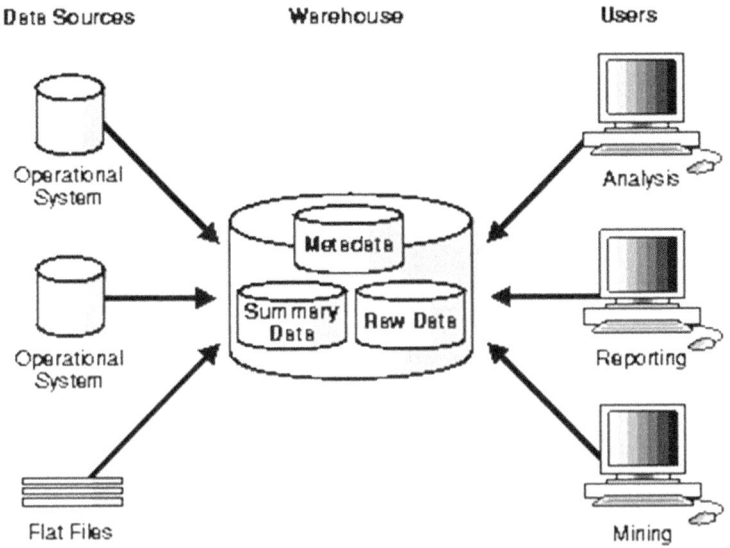

Figure 7.2 Basic Data Warehousing

Basic Architecture: It is a straightforward architecture for a Data Warehouse which facilitates direct access of data, obtained from different data sources. The key components of basic architecture as per Figure 7.2 are as follows:

- Heterogeneous Data Sources
- Data Warehouse
- Users

With a Staging Area: Before putting the data from various data sources into Data Warehouse, the system uses a staging area where the data is cleaned and summarized to make it simpler and finally generate a universal Data Warehouse. The architecture is shown in Figure 7.3.

In the above figure there are 4 components illustrated as mentioned below:

- Data sources
- Staging Area
- Data Warehouse
- Users

With a Staging Area and Data Marts: This Data Warehouse architecture can be customized with different groups for individual organizations. Data Mart can illustrate the storage of purchasing, sales, and inventories separately as shown in Figure 7.4:

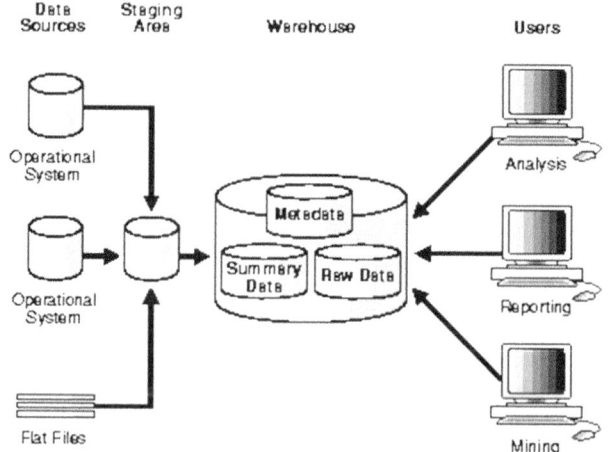

Figure 7.3 Data Warehousing with Staging Area

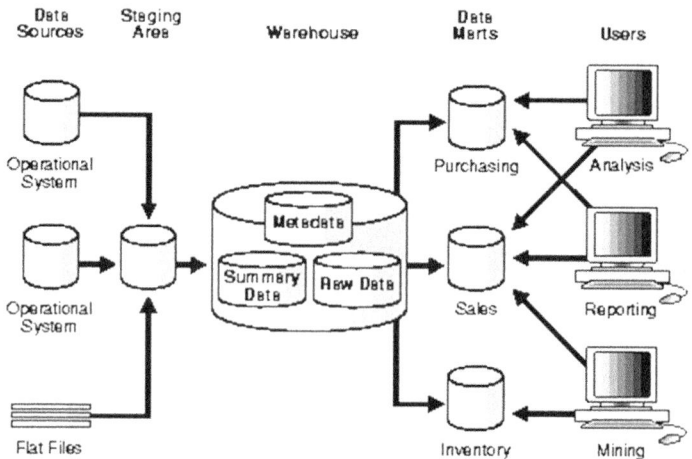

Figure 7.4 Data Warehousing with Data Marts Concept

In the above figure there are five components primarily mentioned:

- Heterogeneous Data Sources
- Staging Area is the area in-between data sources and Data Warehouse.
- Data Warehouse that comprise the metadata, summary data, and raw data.
- Data Marts
- Users

7.4 Benefits of Data Warehousing

Some of the benefits of using Data Warehouse are given below:

- Restructure the data for improving knowledge to business users.
- Restructure the data to deliver excellent query performance without affecting the operational systems for complex analytical queries.
- Support for universal data model.
- Establish single view of data across enterprise.
- Provide consistent data.
- Improved productivity.
- Provide information relevant to organization.
- Lower the computing costs.

7.5 E Governance Model Using Data Warehousing

Project Control in any system is the most significant management task and it ensures timely completion of all the ongoing operations as per plan. Managing properly E Governance project using Data Mining is essential to find out any unpredictable disturbances and their deviation from the expected or planned results. By applying Data Mining to E Governance projects, all progress towards planned performance can be calculated and remedial actions may be initiated for enhanced performance. Figure 7.5 represents E Governance Project Management using Data Mining based Feed Forward and Feedback techniques.

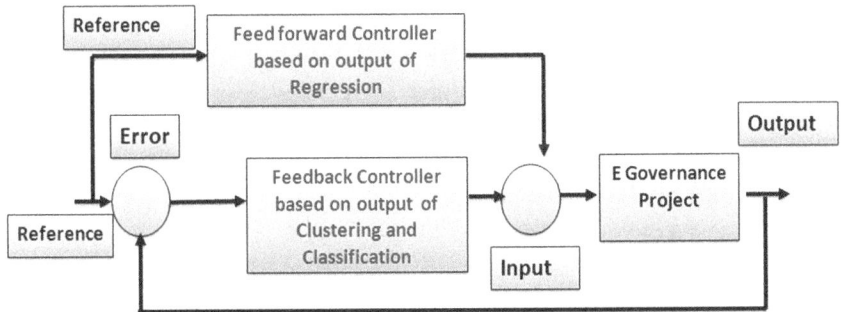

Figure 7.5 E Governance Project Management Using Data Mining Based Feed Forward and Feedback Techniques

In Government organizations, goal, strategic plan, standard of performance and measurements need proper execution of citizen centric services in their Development, Regulatory and Social Welfare Departments [303].

In any kind of project management, performance measurement is a continuous process and can be executed by Clustering, Classification and Regression. Clustering and Classification can be continuously applied within the ongoing E Governance project and its output can be treated as a feedback parameter to adjust any error in the input pattern. Output of Regression can be taken as a feed forward parameter to initiate some proactive approaches within the ongoing project. By using Data Mining based feed forward and feedback techniques, performance analysis becomes easier. The proposed Data Mining based Control System can be useful for every government activity. Here, the projected E Governance model converges all significant aspects of E Governance in one model. The following are four Basic Building Blocks of an E Governance Model.

7.5.1 Module 1: Administration

- ○ E Governance objectives of the proposed Model
- ○ Structure of the proposed System
    - • Centralized Model
    - • Decentralized Model

7.5.2 Module 2: Technical Know How

- ○ Steps for Creating a Data Warehouse
- ○ Effort required for initial setup
- ○ Establishment of State Data Warehouse and State Data Center
- ○ Working of State Data Center
- ○ Data Model Considerations
- ○ Case Study: Data Mining in Department of Education
- ○ Case Study: Data Mining in Department of Health

7.5.3 Module 3: Service Block

- ○ Service Centers
- ○ Self Service
- ○ Kiosk

7.5.4 Module 4: Stakeholder Block

    o Citizen
    o Business

The lowest block as indicated in the figure standardizes the whole government function efficiently for any country. The government is the key enabler and controller in the proposed E Governance Model. Figure 7.6 shows a layered framework of E Governance Model [303].

The overall regulation of government bodies can be accomplished by means of correct technical knowhow. This includes computerization of manual processes, commonly agreed technological standards and quick access databases. The Service Block includes all available operations of the E Governance and provides an interface for government-citizen interaction. The upper block is the Stakeholder Block which includes all concerned parties who deal with the government.

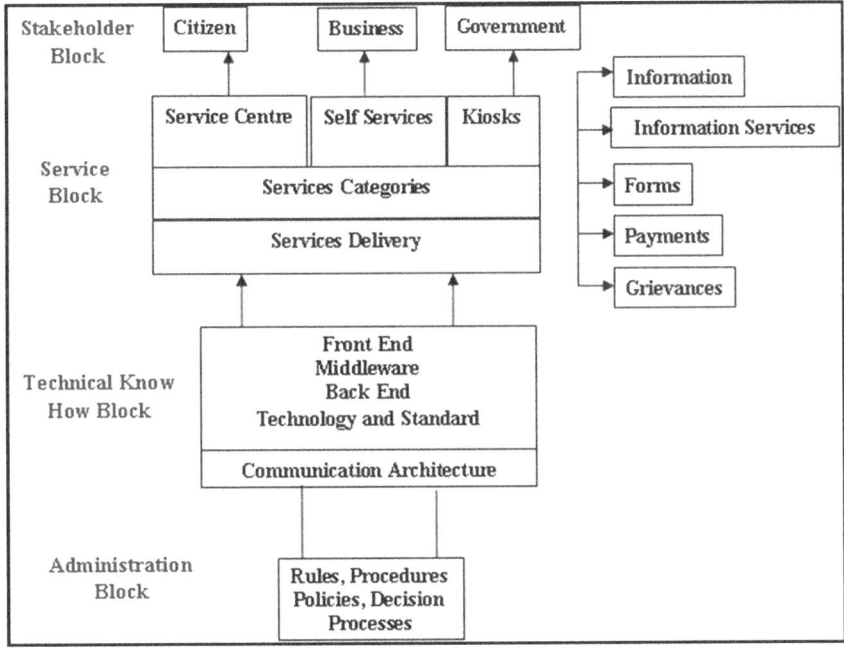

Figure 7.6 Proposed E Governance Model

7.6 Module 1: Administration

The Administration module proposes flow of information and interconnection of different government bodies for E Governance. Any government is framed in hierarchical manner. India for example, has multi tier administration in Central, State and Local Government. The hierarchy of the administration is discussed in Figure 7.7, which needs vertical and horizontal flow of control and information for proper regulation [304].

In independent India, the important feature of administration is the involvement of government in management of resources of the country. To manage all resources at root level, different government bodies play their role with full confidence and control their respective segments. Local Governments are implemented at District, Block, Tehsil and Taluka levels to manage resources of small geographical areas of the country.

Good Governance is the key to full and effective utilization of financial resources and ensures availability, accessibility, efficiency, transparency and accountability. A simple view of transformation of Government is shown in

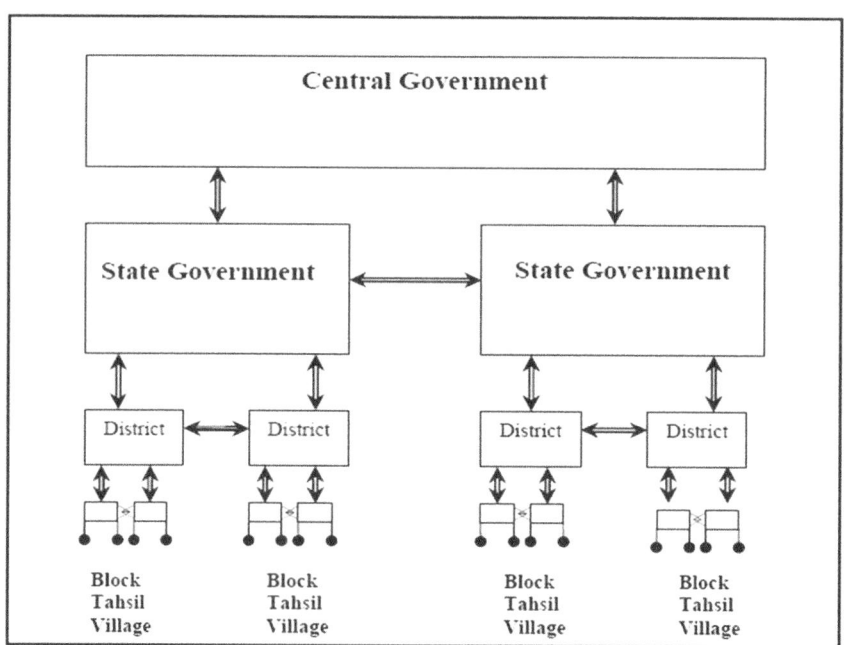

Figure 7.7 Structure of Indian Administration

Figure 7.8. The transition from simple governance to E Governance involves close observation and could be effortlessly understood by the Table 7.1 [305].

7.6.1 Salient Features of the Proposed Model

The sole aim of E Governance is to offer simple functioning practices to government authorities, public and business parties. Information and Communication Technologies can converge all three different sectors to support development and management. Table 7.2 describes significant features of the proposed model.

The features of the proposed model are given below:

- To provide proper information and awareness to the citizens.
- To provide online services and active participation of different citizen services.
- To incorporate ICT in government functions, which provides fast, transparent, answerable, proficient and effective communication with the citizens, businesses and other agencies.
- To decentralize governance results easier decision-making [306].

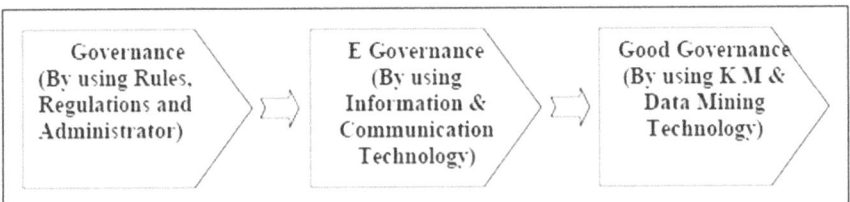

Figure 7.8 Transformation of Government

Table 7.1 Comparison Between Conventional Government and E Government

Goal	Conventional Government	E Government
Vision Direction	Seeking stability	Seeking efficiency and satisfaction
Working Standard	Inside business oriented	Outside customer oriented
Risk Taking	Doina one's best	High quality at low cost
Knowledge Sharing	Less risk taking	More risk taking
Leadership	Limited	Unlimited
Organization Theory	Chief information officer	Chief administration officers
Communication Type	Hierarchy (Single to single point)	Network (Multiple to multiple points)

Table 7.2 Salient Features of the Proposed Model

Introduction to Government	Introduction to Governance
Structure	Functions
Decisions	Processes
Rules	Goals
Roles	Performance
Implementation	Coordination
Outputs	Outcomes
E Government	**E Governance**
Electronic Service Delhery	Electronic Consultation
Electronic workflow	Electronic Controlship
Electronic Voting	Electronic Engagement
Electronic Productivity	Network Societal Guidance

Good Governance

Availability

Accessibility

Efficiency

Traniparency

Accountability

7.6.2 Structure of the Proposed System

The new anticipated model is based on ICT, which may reform organizational structures in both centralized and decentralized manner as described below.

7.6.3 Centralized Model

Centralized E Governance model is represented by single portal for different government services. This model is helpful in reducing cost and resolving integration issues. Centralize E Governance model may share technical, financial and human resources based on single portal system from where a citizen can find all kind of information. Following are the features of Centralized E Governance model.

- All government processes based on ICT is centralized in one organizational unit.
- Limited Infrastructural and set up costs.
- Centralized E Governance models have a single interface for their users and are easy to enforce.

7.6.4 Decentralized Model

In contrast to the above proposed model, Decentralized E Governance model suggests separate implementation of E Governance framework for different departments. Decentralization is helpful in proper decision making, project management and services, funding, revenue collection and operations [307]. Following are the features of decentralized E Governance model.

- All government functions can be distributed among various divisions or organizations.
- High coordination cost and it aims to establish coordination among central and local authorities.

7.6.5 State Level Model of E Governance

The proposed State level model is a mixture of centralized and decentralized approaches. For instance, State Government can be projected as the chief planner and lower government offices with their departments become the partners of the project. Certain important decisions are jointly made and then standardized across the various levels.

- Responsibilities are decentralized at different government departments/levels, with infrastructure and output sharing across the State as a system.
- Generally, high set up costs but more responsive to stakeholder needs. Formation of High level committees to manage various Government activities.

Figure 7.9 describes horizontal and vertical interconnections of E Governance. Intra-department or horizontal and vertical collaborations are extremely crucial for accomplishment of an E Governance project. It is also mandatory to perform governance functions, share information and deliver services to all stakeholders [308].

7.7 Module 2: Technical Know How

For E Governance, there are many applications which need to be automated. Various departments seek computerization and other technological transformation of their working strategies. Now it is necessary to conceptualize a wholesome approach and develop a standard framework and protocols for the regulation of all E Governance activities. The proposed Model utilizes Data Mining and Data Warehousing for better E Governance service performance.

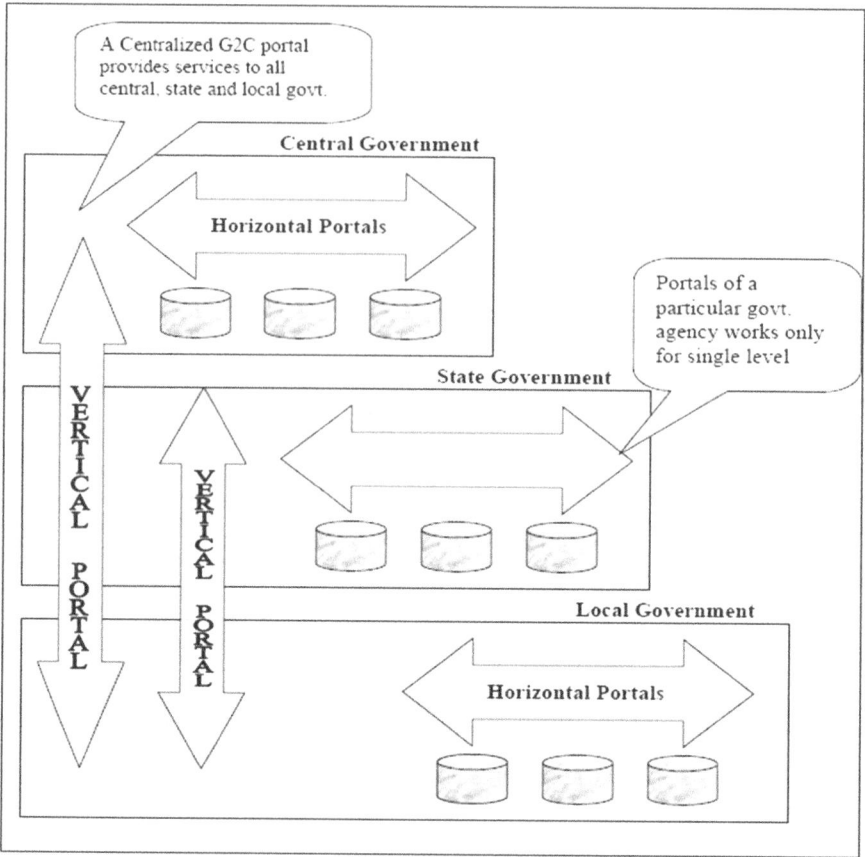

Figure 7.9 Horizontal and Vertical Interconnection for E Governance

7.7.1 Steps for Creating A Data Warehouse

This is one of the significant stages of knowledge discovery operation. A National Data Warehouse is a prerequisite for the proposed E Governance model. In any government system historical data is piled up in form of records and tables. These data sources are in different locations and may use different platforms. Figure 7.10 indicates an integrated framework of data warehouses at national level.

The idea of developing a National Data Warehouse is to combine all distributed database at a central location to provide easy and quick view of all data. The integration of Data Warehouses at multiple levels is indicated in Figure 7.11.

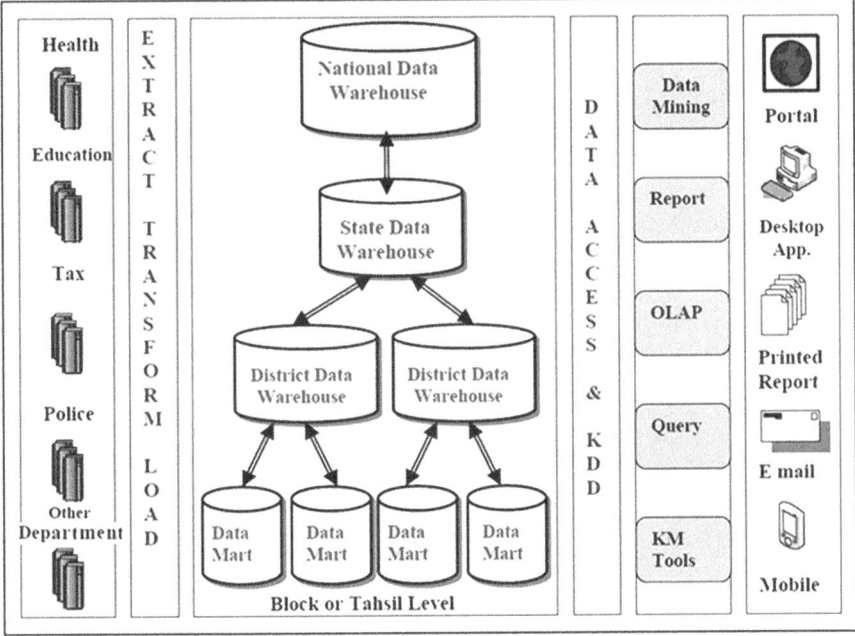

Figure 7.10 A Data Warehouse for E Governance

Different view of data can be employed in different report formats to reflect performance of the working organization. Various report generation and visualization tools are available for different view of data. The proposed Data Warehouse model is a three level system and interconnected together. In reference to the standards of present structure of Indian government, the system should be administered by various Districts, State and Central offices [303].

Thus, Management Information System with Data Warehousing envisages a distributed data processing at District, State and Central levels. As per system specification it is recommended that all the operational records must be submitted to the offices of different departments at District level. After checking its comprehensiveness and readability it should be mailed to the nominated state centers for extraction, transfer and load processing of Data Warehouse. The next step is data consistency test in which data matching is performed as per standards. All state agencies like Ministries of different departments should follow common data consistency standards. The records which successfully pass this stage at the State level are integrated and transferred to the Central Data Warehouse for final complete analysis. Records

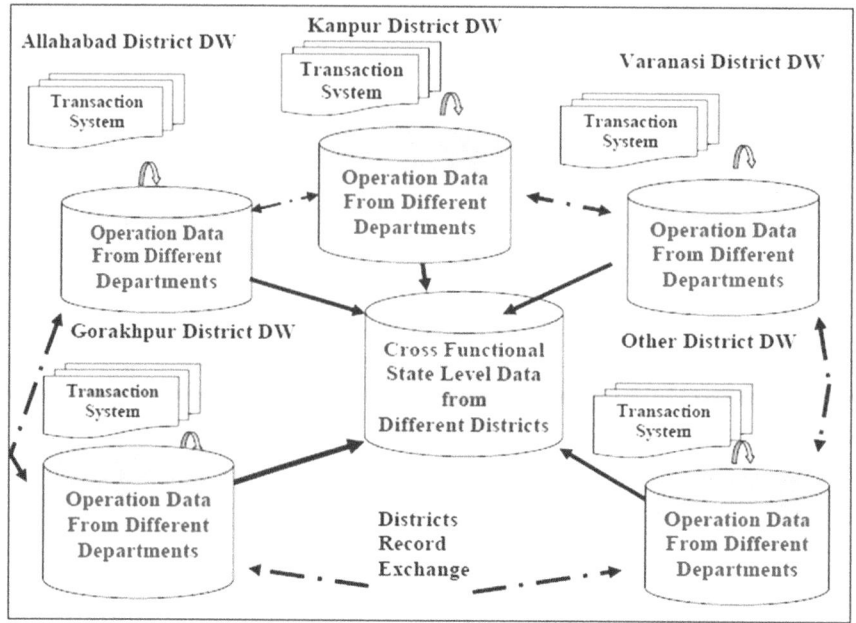

Figure 7.11 Data Warehouse in E Governance

once accepted at Central level are then replicated to the District and State level severs.

7.7.2 Effort Required For Initial Setup of Proposed Model At Different Levels

The tasks and duties of different offices at all the three Districts, State and Central levels have been determined as follows:

7.7.3 District Level

- Filing of paper forms (application, declaration and request forms related to different departments).
- To check data accuracy of the paper based records.
- To offer fundamental services to the community.
- Maintenance of the District data repository.

7.7.4 State Level

- Collection of paper formats from District level
- Data verification and validation of submitted data.
- Transfer of data at Central level
- Maintenance of regional LANs
- Technical assistance to all Districts

7.7.5 Central Level

- Management of Central Data Warehouses.
- Data approval at Central level and its transfer back to the State level.
- Conducting Data Mining analysis task.
- Maintenance of the National level data repository.

7.7.6 Establishment of State Data Warehouse and State Data Center

A State Data Warehouse is needed to be established with a State Data Center. It is a state level data repository including different departmental databases with summarized view. Figure 7.12 explains a state data Center of Uttar Pradesh, India with various departments [303].

7.7.7 State Data Center (Level 3)

- It is a Central Data Warehouse for combined data from all Departments.
- It includes abstract view of the data collected from different Departments.
- It includes only core data elements, which are important as well as unique.

7.7.8 State Level Department Wise Database (Level 2)

- Citizen data from various departments like Regulatory, Developmental and Social Welfare and different Districts are combined at State level.
- Respective State departments will create, maintain and own their combined databases.
- It includes detailed view of departmental databases collected from various Districts.

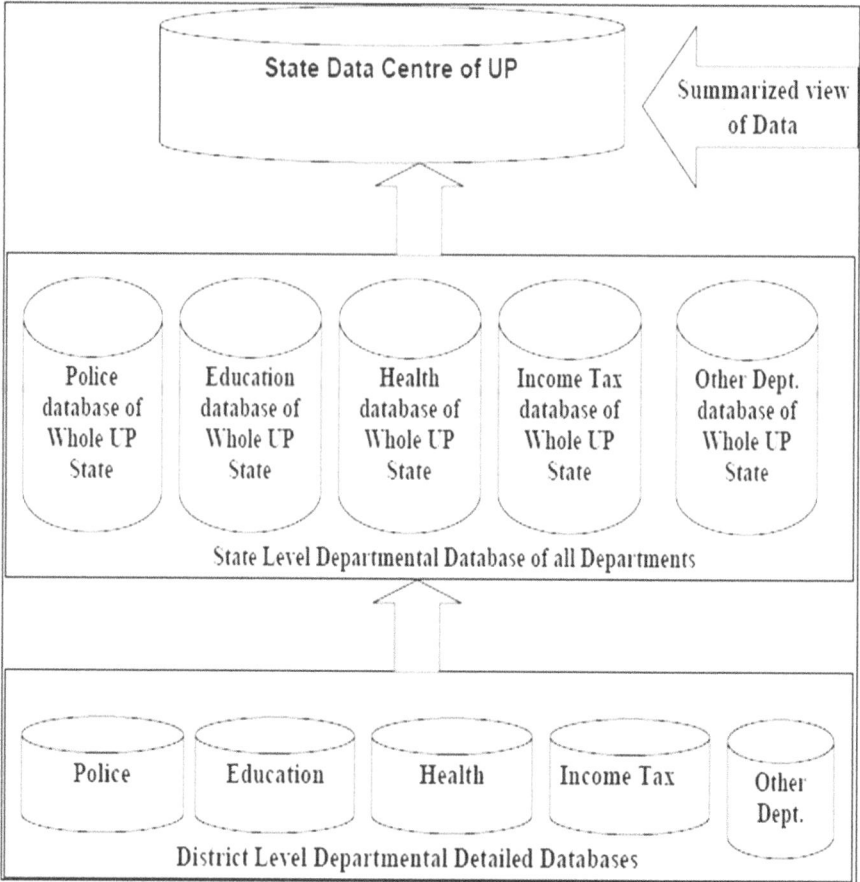

Figure 7.12 State Data Center

7.7.9 District Level Department Wise Database (Level 1)

- At this stage is first combined at the department level.
- It includes detailed department specific data elements collected from different sources of the District.

The Figure 7.12 elaborates that any departmental data is initially collected at District level and then summarized view is forwarded to the state level. Different departments can share data at State level through a common data repository. Figure 7.13 shows the working of transport department as per proposed data levels [303].

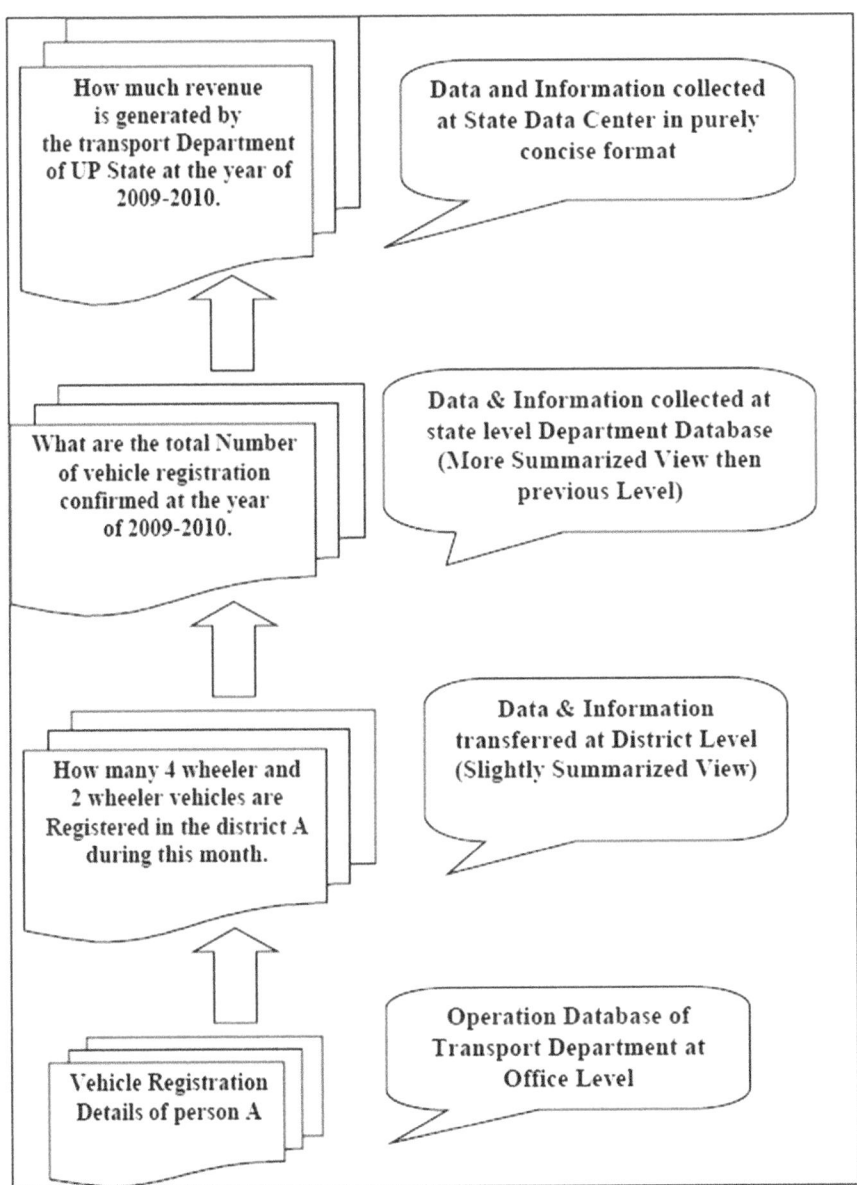

Figure 7.13 Data Transfer Example of Transport Department

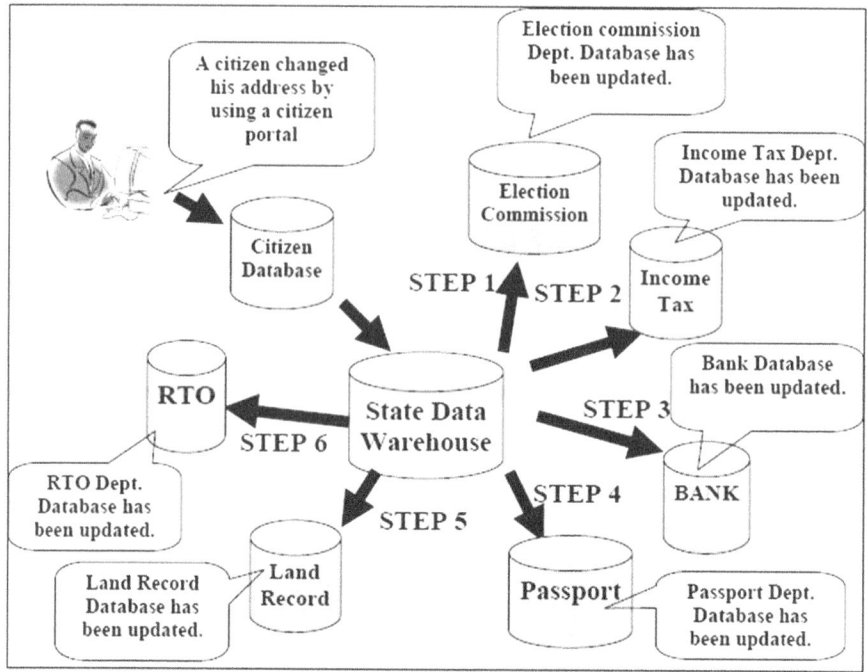

Figure 7.14 Data Updation Through State Data Warehouse

Figure 7.14 shows steps for address updation process related to a person. A single request of address change can be processed by State Data Warehouse and all concerned departments would be updated automatically [303].

7.7.10 Working of State Data Center

The flowchart shown in Figure 7.15 indicates data updation process of a State Data Center. A new record may be initiated through any department and it is matched with existing records. If it does not match with any previously stored record then it may be stored as a new transaction. If matched then it would be updated. This updation may be authenticated by department personnel and then reflected to State Data Center for further reference [303] [308].

7.7.11 Data Model Considerations

There are three types of data taken at different levels [303]:

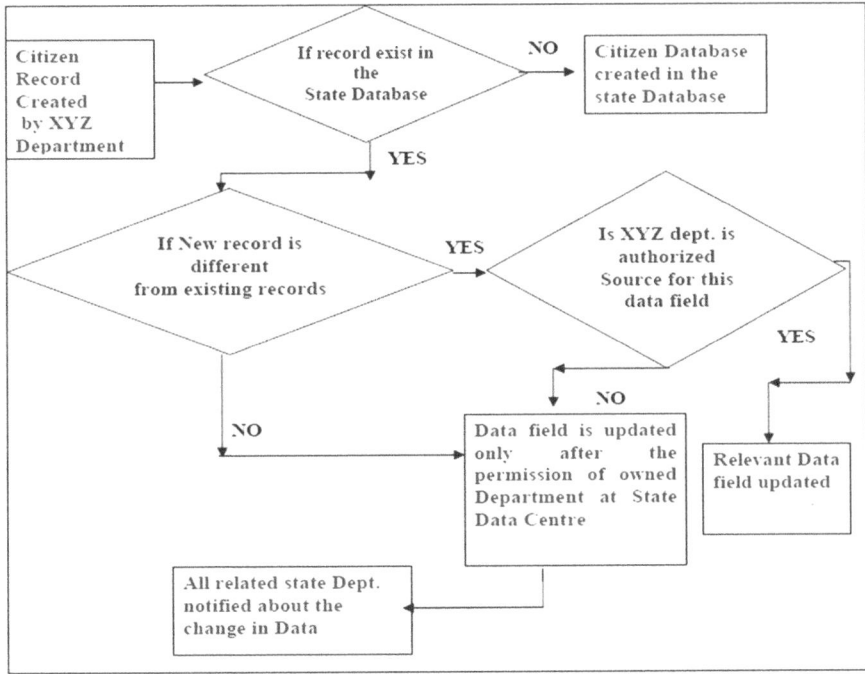

Figure 7.15 Flow Chart of State Data Center

- **Universal Data**: These are common for all departments such as name, PAN number etc.
- **Departmental Data:** These are department level records and needed by other departments such as income tax and bank details.
- **Operational Data:** These are department level transaction records limited to the department and are in large amount.

A State Data Warehouse must include all these data types at its different levels. This data may be created, stored, retrieved and protected under following policies indicated in Table 7.3 [303].

7.7.12 Extract, Transform and Load (ETL)

In this stage domain application data is gathered from different sources. Data normalization and schema development have performed and consequently a repository is developed. This process is known as data staging where the raw operational data are extracted, cleaned, transformed and combined [309]. It

Table 7.3 State Data Warehouse Policy

Data	Data StoragePolicy	Data Retrieval Policy	Data Protection Policy
Universal Data	Data creation and updation by central authority only.	Data updation after proper verification.	Read only format for other departments. Updation only after approval.
Departmental Data	Created by any department as per need.	Only owner department can update	Read only format for other departments. Updation only after approval.
Operational Data	Stored in local databases.	Accessed through Queries.	Limited access at local level only.

Table 7.4 Data Before Cleaning

Name
P.K. Shukla
Mr. R.K.Yadav
Vijay Kumar Singh
Shri Ajeet Ratore M.L.A.

Data Cleaning Process

happens between operation data sources and user databases. Data staging for Extract Transform and Load (ETL) includes following step:

7.7.13 Extract Phase

It is the primary phase of data processing. In this data is selected from various sources from different file formats like relational databases, flat files etc. After data extraction transformation of data could be performed [309].

7.7.14 Data Cleaning

In data cleaning a uniform data representation is followed to establish all its attributes and instances as explained in Table 7.4 and 7.5.

Table 7.5 Data After Cleaning

Title	First Name	Middle Name	Last Name	Suffix
	P	K	Shukla	
Mr.	R	K	Yadav	
	Vijay	Kumar	Singh	
Slur	Ajeet		Rathore	M.L.A.

Table 7.6 Data After Integration

Source 1 Data

Title	First Name	Middle Name	Last Name	Suffix
	P	K	Shukla	
Mr.	R	K	Yadav	
	Vijay	Kumar	Singh	
Shri	Ajeet		Rathore	M.L.A.

Source 2 Data

First Name	Last Name	Age
P	Shukla	27
R	Yadav	59
Vijay	Singh	32
Ajeet	Rathore	60

Title	First Name	Middle Name	Last Name	Suffix	Age
	P	K	Shukla		27
Mr.	R	K	Yadav		59
	Vijay	Kumar	Singh		32
Shri	Ajeet		Rathore	M.L.A.	60

7.7.15 Data Integration

In Data integration data from different sources may be selected and placed at one table as shown in Table 7.6.

7.7.16 Transform Phase

The transform stage contains sets of rules or functions which can generate standard data formats. Some data sources need little manipulation of data which may include Data standardization, Data Enrichment and Data Integration as indicated in Table 7.7, 7.8 and 7.9.

Table 7.7 One Aspect of Data Values

First Name	Middle Name	Last Name	Income per Year ill Rs.
P	K	Shukla	1,50,000
R	K	Yadav	1,90,000
Vijay	Kumar	Singh	3,76,000
Ajeet		Rathore	9,00,000

Table 7.8 Another Aspects of Data Values

Income Range From	Income Range to	Income Tax
1,00,000	1,50,000	10%
1,50,001	2,00,000	20%
2,00,001	Above	30%
2,00,001	Above	30%

Table 7.9 Data Values with Combined Aspects of the Table Above

First Name	Middle Name	Last Name	Income per Year in Rs.	Income Tax
P	K	Shukla	1,50,000	10%
R	K	Yadav	1,90,000	20%
Vijay	Kumar	Singh	3,76,000	30%
Ajeet		Rathore	9,00,000	30%

7.7.17 Load Data

In this process selected data items are loaded into appropriate Data Warehouse.

7.7.18 Data Mart

Data Mart is breaking up of Data Warehouse as small subsets in logical manner constrained to a single field only. For example, a Data Mart of income tax department only contains the income related records of any person and is useful for estimation of patterns in case of proper or improper payment practices. In E Governance, after collection of data from various sources or States, single Data Mart could be prepared for different Regulatory, Developmental and Social Welfare departments at different levels of administration, i.e., State, District, Block and Tehsil level as shown in Figure 7.16 [303].

7.7.19 Distributed Databases

Any government is responsible for performing nationwide operations. To implement Data Warehousing at different level it is necessary to integrate all databases of different departments. Distributed Databases are database related application in which various databases are stored at different departments and in different places.

These databases can be logically integrated by using a central repository. In this arrangement, any department specific data can be accessed from wherever

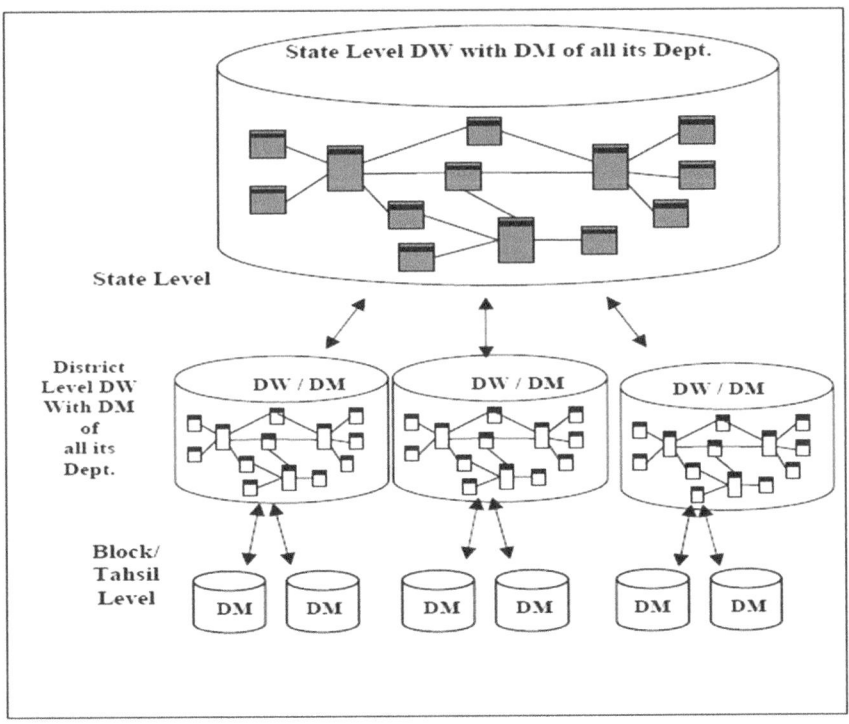

Figure 7.16 State Level Data Warehouse and Data Mart

it is located. In any database related application, distributed database is useful because it minimizes replication of a single database. In any information system all departments can utilize distributed database efficiently. Such kinds of arrangements are useful in terms of minimization of efforts and duplication. The Figure 7.17 shows Data Mining in different government departments by using distributed database model.

7.8 Module 3 Service Block

In the service block, services of E Governance are provided to the citizens for their betterment. It provides an interface so that a common citizen can also participate in decision making processes. The Service Block is also helpful in simplifying complex government processes involving too many offices and man [310] [311]. The service block facilitates information access, making

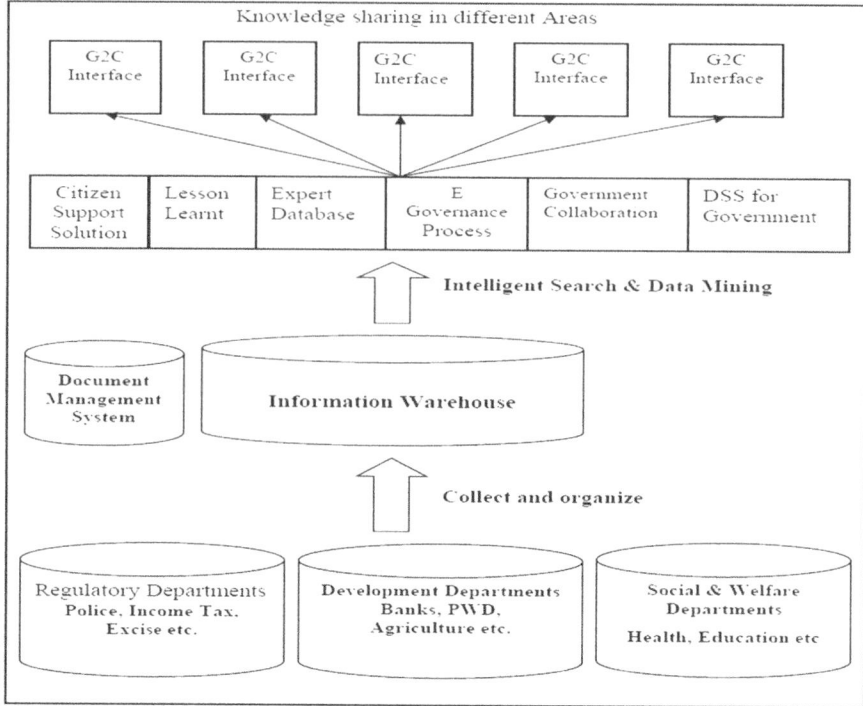

Figure 7.17 Data Mining in Different Government Departments by Using Distributed Databases

payments, submitting complaints and downloading forms. Following are the services offered by the proposed model:

7.8.1 Service Centers

Service centers are the front-end delivery points. The aim of service center is to facilitate services for rural citizens. The Service centers may work on the behalf of Government, Private organizations and NGOs. In these services centers various essential services are integrated to fulfill the requirement of rural citizens from one point. It is capable of doing so by using Information Technology enabled infrastructure or without using it. The aim of the service centers is to encourage rural entrepreneurship, full utilization of rural resources and increased employment, improved society responsiveness for social revolution [312].

7.8.2 Self Service

A self-service approach encourages citizens to help themselves at any time anywhere by means of web enabled services. Many departments now provide a number of their services by means of their websites and in form of information such as reports, forms, etc. by using web based applications. This reduces efforts, time and money for all the concerned.

7.8.3 Kiosk

Kiosks are E Governance counters, installed at various public places. Government has planned to launch kiosks across the all States so that citizens in remote locations can also access the Government-Citizen (G2C) interface [313].

7.9 Module 4 Stakeholder Block

Stakeholder can be an individual, a group of persons or a community with common area of interest. The most important groups are identified in 3 parts:

7.9.1 Citizen

Citizen is associated to E Governance by using Government to Citizen (G2C) interface. G2C interface enables online interaction between government and private individuals. With the establishment of G2C the Government sectors become more approachable to the citizens. It also offers correct, timely and reliable information at one place. Citizen Service Centers (CSCs) are the example of G2C interface established by government at every village panchayat level. The citizen may visit the center to use various services such as date of birth registration etc [312].

7.9.2 Business

Business is associated to E Governance by using Government to Business (G2B) interface. G2B interface enables online interaction between business groups and government parties, i.e., Local and Central, in order to perform various trades and business related transactions [314].

7.9.3 Government

Various government departments are associated to one another by means of E Governance using Government to Government (G2G) interface to facilitate online interaction at different levels of government. The interaction is helpful in developing new relations among different departments of the government [314].

7.10 Conclusion

The chapter discussed about basics of Data Warehousing and its application in E Governance. In this chapter an E Governance model framework based on Data Mining and Data Warehousing techniques which may be efficiently used by the Government at all its administrative levels (National/State/District/Block) is also discussed. The proposed Model is designed to serve all possible aspects of E Governance with the help of four basic building blocks: Administrative Block, Technical Know How Block, Service Block and Stakeholder Block. A Data warehouse model is presented which is a three level system and is interconnected. It has a State Data Center which is primarily a data repository including different departmental databases. To explain a typical State Data Center, the case of Uttar Pradesh with its different departments has been discussed. It is important to check data consistency within a State Data Center for which a standard data consistency model has been proposed. The research outcomes have also been qualitatively described for the usefulness of government in terms of proper decision making and establishment of availability, accessibility, efficiency, accountability and transparency for the citizen of the nation.

8

Government Data Mining Case Studies on Education and Health

8.1 Introduction

Holistic development of a nation constitutes, at the very grass root level, two major areas needing reforms namely Education and Health. Both the Central and the States' governments have laid great emphasis on these and have inculcated projects like "Education for All" and "Pulse Polio Immunization". With the objective to analyze the efficiency of such ongoing government projects, the present investigation has obtained an education dataset from "Education for All" and Polio dataset from "Pulse Polio Immunization" projects thereby examining status of primary education and Polio immunization in Uttar Pradesh using Data Mining techniques [315][316].

In this chapter, Clustering, Classification and Regression, the three most useful Data Mining tools have been developed using MATLAB 7.9 with user friendly interface to explore interesting patterns which could be used in future policy establishments at State and National level. Also, an open source tool WEKA has been used for comparison purpose. The algorithmic performances of all above approaches have been established under specific criteria and their merits have been presented. The research outcomes have also been qualitatively described for the usefulness of government in terms of proper decision making and establishment of availability, accessibility, efficiency, accountability and transparency for the citizen of the nation.

8.2 SVM Based Data Classification and Regression Model for E Governance

Initially A classification based model is developed that includes Data Preprocessing, types of SVM classifiers along with different types Kernel functions [317]. Data Preprocessing is important because E Governance projects

E Governance Data Center, Data Warehousing and Data Mining: Vision to Realities, 155–202.

generally consist of huge datasets, which needs considerable effort in Data Cleaning. For SVM implementation four components have been developed. The first component is simply an implementation of traditional SVM approach. In improvement over this by using a weight function is represented as second component. The third component includes a traditional LS-SVM implementation and lastly the same weight function used in this also and represented as component four. The Figure 8.1 represents the proposed Data Mining techniques for E Governance in which the blocks having dotted lines are indicating the novel techniques developed here. Regression is an efficient

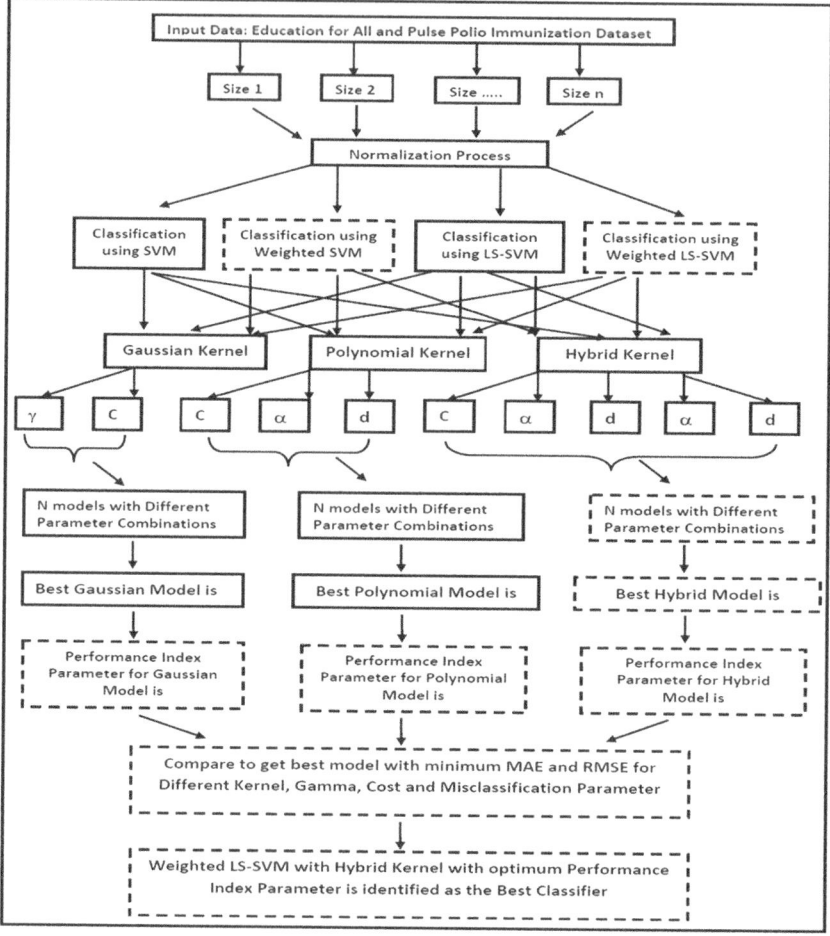

Figure 8.1 Work Flow Diagram of Proposed Classification Techniques

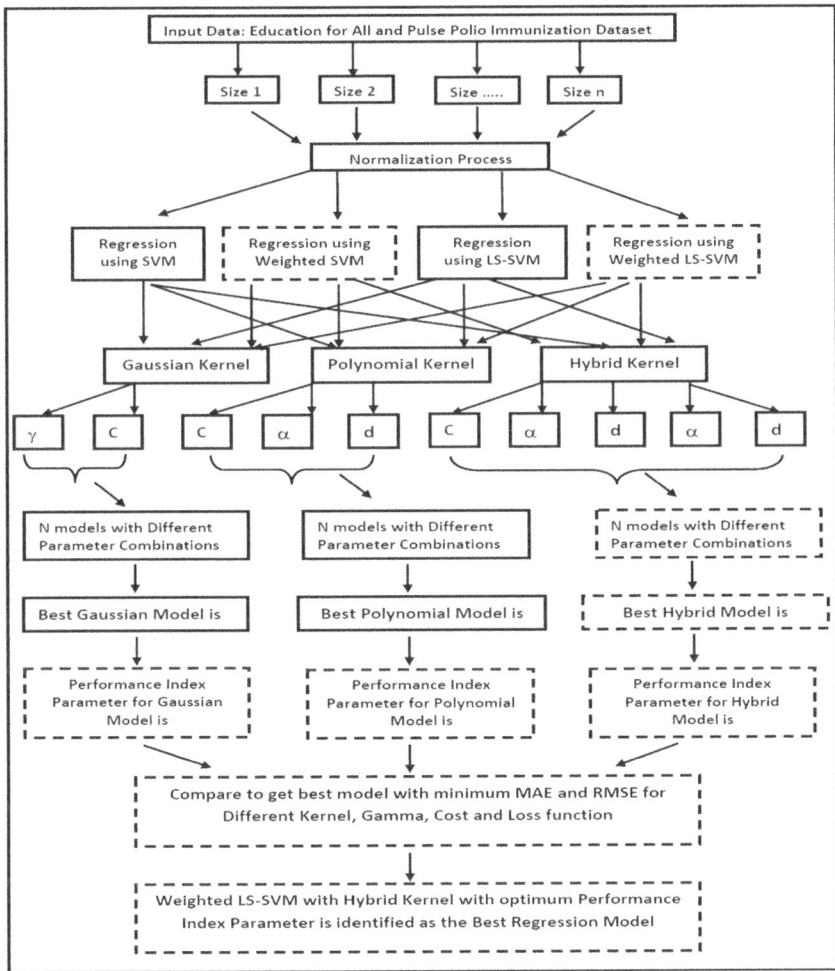

Figure 8.2 Work Flow Diagram of Proposed Regression Techniques

way to predict future outcomes based on previous trends. In this investigation an E Governance based Regression model is also proposed, which is very much similar to Classification.

The Figure 8.2 shows work flow diagram of proposed Regression techniques with a special loss function. A newly developed Kernel function has been developed with enhanced features. The performance of modeling methods is depends upon choice of proper or optimal hyper parameters, which is difficult to choose. Hence, this model provides a convenient way of model parameters selection for all available components in order to establish the best combination of parameters.

A Performance index Parameter (PIP) function has been developed to check the performance of individual classifiers on the basis of standard errors. There are total 9 error terms are defined and either all or any of the error may be calculated for any classifier with all kinds of Kernel functions and a least error classification outcome may be considered for further result interpretation. This investigation focuses on Mean Absolute Error and Root Mean Square Error for establishment of best pair of classifier and kernel.

Since E Governance problems are real time with diverse datasets and neither of the available SVM and LS-SVM implementations fulfills the criteria above, the proposed model provides a sparse, robust, easy to construct and computationally efficient solution by extending SVM and LS-SVM in terms of Hybrid Kernel and Weighted Support Vector model [318] [319]. This model is extremely helpful for real time data analysis, less complex and can easily adapt different datasets especially with the presence of noisy data. This novel approach has been combines all desirable features all together for better E Governance Project Management.

8.2.1 Data Preprocessing Techniques for the Proposed Model

Data preprocessing is a method to clean and transform the data before any data modeling operation. Data cleaning is useful for removing the noise and outliers. Here missing data is treated manually and replaced by most likely probable values [320]. In this research work dimension reduction is required with "Education for All" datasets because it contains extremely detailed parameters as its attributes. Normalization is a type of data transformation for improving the classification accuracy and subsequently performance of Data Mining involving Support Vector Machines and Fuzzy C Means Clustering. Normalization process provides better results if it is scaled properly. In this work data is normalized between the range of [4, -4] for SVM classification and between 0 to 1 for Fuzzy Clustering. It is also required to convert datasets from categorical format to nominal format such as Male and Female category is represented as 0 and 1. For classification task data is transformed into ordinal format.

8.2.2 SVM Kernel functions for the proposed Model

Data classification could be differed under two conditions. The easiest classification could be done when data points are linearly separable but most of real datasets are complex and linearly inseparable. A nonlinear distribution of data points are transformed into linearly separable case using Support Vector Machines. If non linear data points are plotted on higher dimension plane a clear linear separation may be achieved. This could be possible by a special function known as Kernel function. When a suitable Kernel function is applied to the input data it generates a high dimension feature space which is having linear separation [321]. This may enhance classification accuracy up to major extent.

8.2.3 Properties of Kernels

In any Kernel X can be represented as input space for given dataset. The inner product feature space is represented by F. φ is a mapping function converts input space into feature space. A Kernel function is shown as a function of K and here for all values of x:

$$z \in X, \, K(x, z) = \phi(x).\phi(y)$$

K is a Kernel function if $K : X \times X \to R$

Suppose, K_1 and K_2 are Kernel functions then all linear combination of K_1 and K_2 must be considered as a valid Kernel. For example:

1) $K_1 + K_2$ is a kernel function
2) $ak_1, a > 0$ is a kernel function

Here, it is preferred to have SVM classification with maximum accuracy when applied with suitable Kernel functions [321].

8.2.4 Types of Kernel

Following are the Kernel functions applicable for Support Vectors Machines as shown in Figure 8.3 [321] [322].

Two generic representations of Kernels are local Kernels and global Kernels. Local Kernels use a distance function for the generation of high dimension feature space. All dense data points may be considered for generation of high dimension feature space. Gaussian Kernel is the most common local Kernel type. Global Kernel usage dot product representation and it consider all type of data points under some degree for the generation of feature space. All common Kernels such as Linear, Polynomial and Sigmoid are the type of global Kernel. The performance of SVM approach is very much depending on Kernel functions. All existing Kernels have their own benefits and shortcomings.

Kernel Types	Formula
Linear	$K(x,y) = x \cdot y$
Polynomial	$K(x,y) = (1 + x \cdot y)^d$
Sigmoid	$K(x,y) = \tanh(\alpha \cdot x \cdot y + \beta)$
Exponential RBF	$K(x,y) = \exp(-\gamma \|x - y\|)$
Gaussian RBF	$K(x,y) = \exp(-\gamma \|x - y\|^2)$
Multi-quadratic	$K(x,y) = -\sqrt{\|x - y\|^2 + c^2}$

Figure 8.3 Types of Kernel

8.2.5 Hybrid Kernel

Local and global both can perform well if chosen for suitable datasets. But it is complicated to find out an efficient matching between Kernel function and suitable datasets. A combination of local and global Kernel may be proven as best option to enhance the classifier performance. This may be termed as Hybrid SVM Kernel. There is no well proven theory available about the mixing of Kernels. This could be only motivated on the basis of end result.

In this present investigation initially the performance of individual Kernel is initially tested with chosen datasets. Then a fusion of existing Kernel functions has been derived to incorporate the advantages of different Kernel in a single Kernel functions. Polynomial and Gaussian Kernel are prominent Kernel functions influence the execution of Support Vector Machines. When Polynomial Kernel is tested with SVM model its generalization performance is outstanding but learning ability is poor. While in case of Gaussian Kernel generalization is very poor. A fusion of Polynomial and Gaussian Kernel has been proposed to boost the model performance. The Hybrid Kernel function developed and performance of the model is tested by using "Education for All" dataset and Polio dataset. Following is the expression for Hybrid Kernel.

$$\mathcal{H} = Z * \mathcal{P} + (1\text{-}Z) * \mathcal{R}$$

H = Hybrid Kernel
Z = A Scalar Quantity (between 0.1 to 0.9)
P = Polynomial Kernel

8.2.6 Weighted Support Vector Machines with Hybrid Kernel for the Proposed Model

Although SVM Classification and Regression are promising techniques but their results may be data dependent. In any dataset attributes plays a major role in model estimation and prediction. It may be possible that some attributes have more importance as compared to others. The attributes may have instances with small variance and its values may play major role in model estimation and prediction. The belongingness of instances in any particular class may be uneven and this may affect the classification accuracy. If there is a possibility that random errors have uneven or irregular variance then some measures has to taken to adjust the error.

Here, it is proposed that appropriate weights have been introduced to suppress data variance error. The error term is as follow:

$$s = \sum_{i=1}^{N} w_i e_i^2 = \sum_{i=1}^{N} w_i (d_i - y_i)^2$$

The weight function is outlined as follow:

$$A^T WAu = A^T Wv$$

Matrix of weighted coefficient is shown below:

W= diag$\{w_{1.........}w_N\}$

The weight proposed in above expression is capable enough to regulate the quantity of influence every records have on the model estimation and prediction. Here, in this proposed approach the weight has been calculated by the dataset. The attribute having significant impact for model estimation has been used to derive the weight coefficient. The value of weighted coefficient is supposed to be high if the data points have less noise and best fit and it should be assigned minimum to outliers. So in such way it could minimize the variance of data points and consequently error caused by this variance. Similarly weight can also be estimated as per its distance from the hyper plane. The data points closer to fitted line may be assigned to more weight as compared to the data points located far from the fitted line [323][324][325].

8.2.7 Performance Index Parameter (PIP)

The Performance Index Parameter (PIP) is proposed to describe the efficacy of the anticipated model for a set of observations. The Performance Index

[pip] = pipfit2(m.y)

[pip] = pipfit2(m.y.gFitMeasure)

[pip] = pipfit2(t.y.pipFitMeasure.options)

Performance Index Parameter	Optimization parameters
'all'	calculates all the measures below
'1'	mean squared error (mse)
'2'	normalized mean squared error (nmse)
'3'	root mean squared error (rmse)
'4'	normalized root mean squared error (nrmse)
'5'	mean absolute error (mae)
'6'	mean absolute relative error (mare)
'7'	coefficient of correlation (r)
'8'	coefficient of determination (d)
'9'	coefficient of efficiency (e)
'10'	maximum absolute error
'11'	maximum absolute relative error

Figure 8.4 Performance Index Parameter

Parameter (PIP) summarizes the inconsistency between recorded values and the values expected in the proposed model. The Performance Index Parameter (PIP) is defined as follows:

Here m represents the vector of target values for proposed model. y represents matrix or vector of output from proposed model and pipFitMeasure is a array of string values representing different form of Performance Index Parameter (PIP) measures as shown in Figure 8.4.

All the parameters have been important to quantify the performance of Classification and Regression. In the present investigation following parameters have been viewed for performance analysis [326].

Mean Absolute Error: $MAE = \frac{1}{n} \sum_{i=1}^{n} |f_i - y_i| = \frac{1}{n} \sum_{i=1}^{n} |e_i|.$

Root Mean Squared Error: RMSE = $\sqrt{(\sum_{i=1}^{n}(f_i - y_i))/n}$

f_i: Predicted value by model.
y_i: Total values observed.
n = Data points.

8.2.8 Support Vector Machines Classification Using Hybrid Kernel

Figure 8.5 shows the flow chart of SVM Classification which includes steps of model parameter selection, kernel type selection, training as well as testing. Following are the steps of SVM Classification [327].

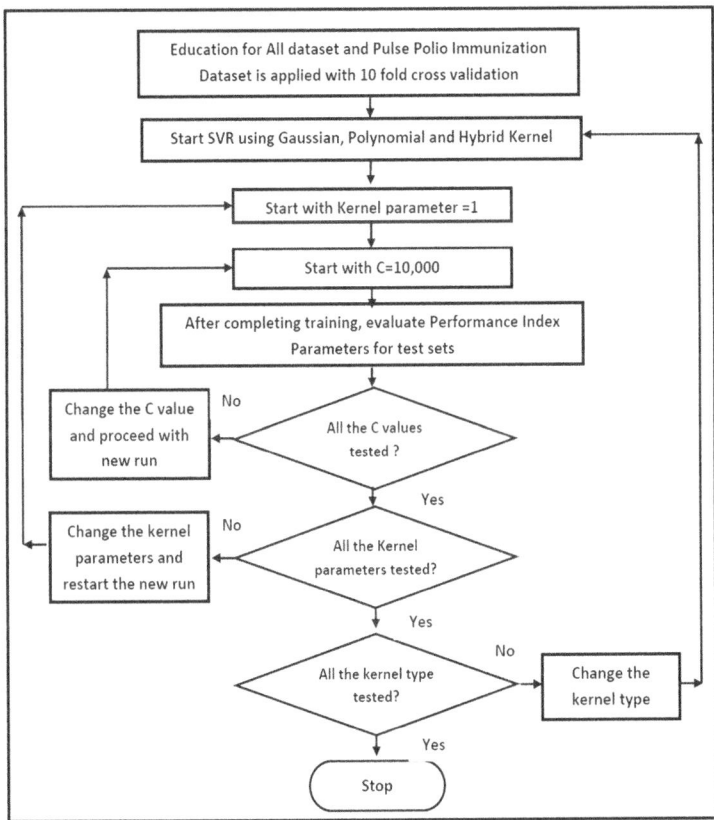

Figure 8.5 Flow Chart for SVM Classification

Step 1: Load the dataset in excel format and take classes as input to SVM classification in order to train the model

Step 2: Initialize the input parameters such as:

- Kernel (Polynomial, Gaussian and Hybrid)
- Sigma
- Misclassification cost

Step 3: Train the model.

Step 6: Perform cross-validation to have best C , Kernel Parameter and Kernel type also calculate Classification errors by using Performance Index Parameters.

Step 7: Calculate Support Vectors and predict the class using input data and support vector.

Step 8: Plot data-samples, hyper-plane and support vectors.

Step 8: Test the Model and check the model using Performance Index Parameter (PIP).

8.2.9 Weighted Support Vector Classification with Hybrid Kernel

In this Weighted SVM, a weight function imposed in each data model as per data attributes has been introduced to sharpen the performance.

8.2.10 Least Square Support Vector Machines with Hybrid Kernel

Like SVM, traditional Least Square Support Vector Machine (LS-SVM) is also modified by using Hybrid function to incorporate the advantages of both Polynomial and Gaussian Kernel. It is important to verify the performance of each kernel so that choice of appropriate Kernel can be made for further analysis. Following are the steps for LS-SVM [328].

Step 1: Load the dataset in excel format and Initialize the input parameters such as:

- Bound on lagrangian multipliers.
- Hybrid Kernel and its parameters.

Step 2: Apply LS-SVM classification.

- First transform the non linear data points into higher dimension space so that separation becomes linear easily using kernel mapping.

- Then in order to find out hyper-plane solve the quadratic equation and find out the lagrangian multiplier and objective function.

Step 3: The outputs of SVM Classification function are support vector, number of support vector, position of support vector, objective function, bias, weight, lagrangian multiplier and time for processing scalar product.

Step 4: Then call svmval function to predict the class using input data and support vector.

Step 5: Plot data-samples, hyper-plane and support vectors.

8.2.11 Weighted Least Square Support Vector Machines with Hybrid Kernel

LS-SVM is an implementation of SVM with less complexity and better performance. In this approach it transforms the quadratic equations as linear equation for model generation. Although the performance is better in LS-SVM but still it is suffering with some inaccuracies during classification. Therefore, here again a weighted LS-SVM has been proposed. Herein a weight parameter could be used to eliminate abnormality of the data sample by considering crucial data more significantly. Following are the steps of WLS-SVM.

Step 1: Load the dataset in excel format and Initialize the input parameters such as:

- Bound on lagrangian multipliers.
- Hybrid Kernel and its parameters.

Step 2: Apply WLS-SVM classification.

- To introduce a weight, it is required to choose the attributes which are more crucial in terms of decision making. That attributes should be given more weightage and than transformed into high dimension space using Hybrid Kernel in order to have linear separation.
- Then in order to find out hyper-plane solve the quadratic equation and find out the lagrangian multiplier and objective function.

Step 3: The outputs of WLS-SVM-classification function are support vector, number of support vector, position of support vector, objective function, bias, weight, lagrangian multiplier and time for processing scalar product.

Step 4: Then call svmval function to predict the class using input data and support vector.

Step 5: Plot data-samples, hyper-plane and support vectors.

8.2.12 Support Vector Regression with Hybrid Kernel

Support Vector Machines are capable enough to perform both pattern recognition as well as Regression. In Support Vector Regression, it is intended to get a regression curve using given samples [328]. Regression can also be performed linear and non linear datasets. A loss function is introduced to generate robust and sparse model. The sparse model is used for dependency assessment of dataset in feature space. Numbers of support vectors are controlled by a special variable. The steps of Support Vector Regression have been explained by using Figure 8.6.

Step 1: Load the dataset in excel format and take classes as input to SVM classification in order to train the model.

Step 2: Initialize the input parameters such as:

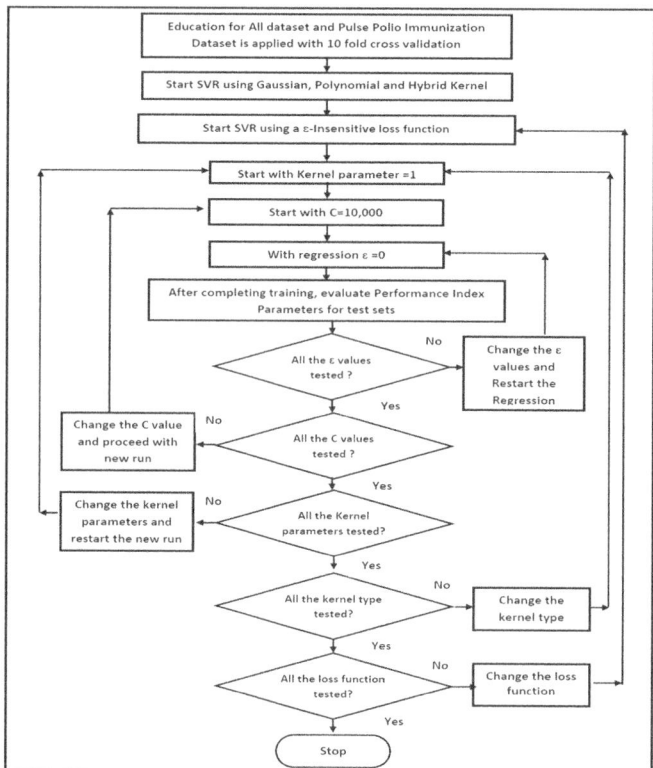

Figure 8.6 Flow Chart for Support Vector Regression

- Kernel (Polynomial, Gaussian and Hybrid)
- ε- Insenstive loss function
- Kernel Paramters

Step 3: Train the model.

Step 4: Perform cross-validation to have best ε, C , Kernel Parameter and Kernel type and also calculate Classification errors by using Performance Index Parameters.

Step 5: Calculate Support Vectors and predict the class using input data and support vector.

Step 6: Plot data-samples, hyper-plane and support vectors.

8.2.13 Weighted Support Vector Regression with Hybrid Kernel

In this Weighted SVR, a weight function imposed in each data model as per data attributes has been introduced to sharpen the performance.

8.2.14 Least Square Support Vector Regression with Hybrid Kernel

LS-SVR is a simple approach having important parameters as Kernel settings and γ. The effectiveness of the model generally depends upon these two parameters. Choosing appropriate values for Kernel function and γ is important to get optimal Regression model. There is no established approach available for this. The exclusive technique is to apply cross validation or leave one out approach and check the model activities in terms of prediction error for different settings. It is essential to choose model with smallest prediction error. This could be done by using Hybrid Kernel. Once optimal parameters have been identified, training and testing have been completed.

8.2.15 Weighted Least Square Support Vector Regression with Hybrid Kernel

This approach is includes similar steps of LS-SVR with an inclusion of a weight function as discussed earlier to enhance the good fitting of the model and minimize the complexity.

8.3 Fuzzy C Means Clustering Model for E Governance

In E Governance Fuzzy C Means Clustering is important because hard boundaries for the separation of different groups are difficult to determine. In Fuzzy

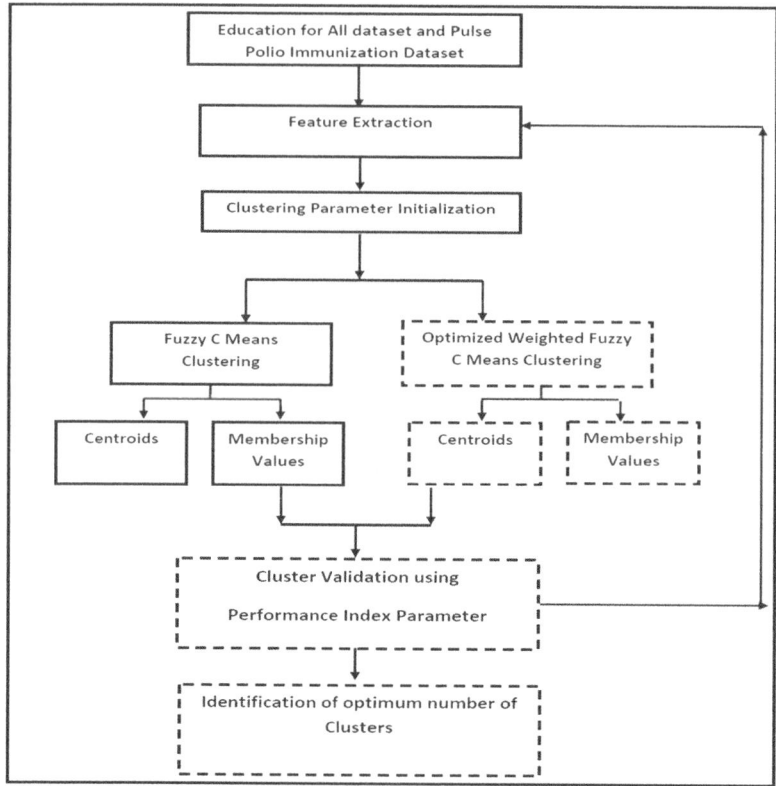

Figure 8.7 Work Flow Diagram of Proposed Clustering Techniques

C Means a data point has membership in more than one cluster with some degree of membership [329]. The Figure 8.7 illustrates two possible modules of fuzzy clustering. In module one traditional Fuzzy C Means has been implemented and in module two, an improvement over traditional fuzzy C Means is implemented. The performance of the clustering for both of these cases is compared by using Performance Index Parameters. This Performance index Parameter should have optimum values in order to determine most suitable number of clusters.

8.3.1 Steps for Fuzzy C Means Clustering

Following are the steps of FCM clustering [329].

Step 1: Load the dataset in excel format and Initialize the input parameters such as:

- Initial Membership Matrix
- Error constraints
- Termination measures
- m=2
- No. of Clusters
- No. Iterations.

Step 2: Perform Fuzzy C Means Clustering and Check the condition if error is greater than tolerance then

- Update membership matrix U
- Calculate Center v_i
- Calculate Distance d_{ik}
- Calculate Membership μ_{ik}
- error=$\left\| \mu_{ik}^{new} - \mu_{ik}^{old} \right\|$

Step 3: Test the Model and check the model by using the Performance Index Parameters and on the basis of best set of parameters choose the optimum number of clusters.

Step 4: Take the Best value for number of cluster and plot clusters.

8.3.2 Steps for Optimized Weighted Fuzzy C Means Algorithm (OWFCM)

The important parameters of Fuzzy C Means are distance function, cluster Center and degree of fuzziness. Following are modifications have been introduced in traditional method as Optimized Weighted Fuzzy C Means Algorithm:

- **Modification of Distance:** Traditionally mean distance has been used for degree of membership calculation. Here, we are using "Harmean" distance which is expected to have more accurate degree of membership.
- **Weighted Cluster Centre:** Weighted cluster centre is calculated by following formula:

$$f_0 = (d./(sum(d, 2) * ones(1, c)));$$
$$Where d = d = (d + 1e - 10).^\wedge (-m/(m - 1));$$

Where eta is typicality weight, which default value is 4. For estimating the centroids, the weighted clusters are used for suppressing the undesirable effect of outliers.

- **Weighting Exponent m:** The m is defined as amount of fuzziness and it represents weighting exponent for each fuzzy membership it is recommended that m should always greater than one.

Following are the steps of Optimized Weighted Fuzzy C Means Algorithm (OWFCM).

Step 1: Load the dataset in excel format and Initialize the input parameters such as:

- Initial Membership Matrix
- Error constraints
- Termination measures
- M > 1
- No. of Clusters
- No. Iterations.

Step 2: Perform Optimized Weighted Fuzzy C Means Clustering and Check the condition if error is greater than tolerance then

- Update membership matrix U
- Calculate **weighted cluster Center v_i**
- Calculate of **modified Distance d_{ik}**
- Calculate Membership μ_{ik}
- error= $\left\| \mu_{ik}^{new} - \mu_{ik}^{old} \right\|$

Step 3: Test the Model and check the model by using the Performance Index Parameters and on the basis of best set of parameters choose the optimum number of clusters.

Step 4: Take the Best value for number of cluster and plot clusters.

8.4   Data Mining Tools

Data Mining is a growing technique with several existing commercial tools. Different tools have different parametric requirements and performance. The thesis work chooses WEKA because it is open source and even work with small data sets.

8.4.1 WEKA

WEKA, a software Data Mining tool owned by University of Waikato based in New Zealand. WEKA serves as an implementation tool for Data Mining

Operations. It is developed in Java and includes collection of Java Classes to perform various Data Mining tasks. WEKA also offers convenient methods of Data preprocessing and Data post processing. Effective visualization of Data Mining analysis is also available with the WEKA System. Using WEKA system we can process the data set, provide input to the learning system and examine the performance. WEKA software provides various Data Mining techniques with easy interface. It is already mentioned that WEKA uses various Java packages to perform Data Mining operations. Each Java package consists of a single machine learning algorithm. Any Data Mining algorithm could utilize one or more Java packages to perform the complete Data Mining analysis. The Java classes can be called by using command line interface. There is variety of options and commands available to control the input and output. Help option is also there for finding help messages. Following are the details of important packages used in WEKA System [330][331].

WEKA.Core: It is main part of the WEKA System contains classes that can be referred by all packages. It provides all basic supports regarding identification of valid data sources, calculation of standard functions and basic data structure needed by the other packages.

WEKA.Filters: Filtering is important for identification of useful attributes from data sets. The Filter package includes a group of classes which are useful for selection of data items during data preprocessing. Data preprocessing is an important task and it includes data cleaning, attribute selection, normalization and data sampling.

WEKA.Classifiers: The role of the WEKA Classification package is to perform classifies and predicts methods of Data Mining. Here, WEKA.classifiers is the main class and supported by some other classes.

WEKA.Clusterers: The package is responsible for implementing clustering algorithm of Data Mining. In this algorithm clusters are recognized as per set parameters.

WEKA.Associations: This package contains algorithm which is responsible for association rule finding.

WEKA.GUI: This package includes classes which are responsible to develop Graphical User Interface. In WEKA important GUI tools are Arff file viewer, Simple Command Line Interface tool, Explorer, Experimenter and the Knowledge flow tool. The Result Visualization tool is also a part of WEKA GUI package.

Other Packages: Following are the supportive packages which are utilized by the package mentioned above:

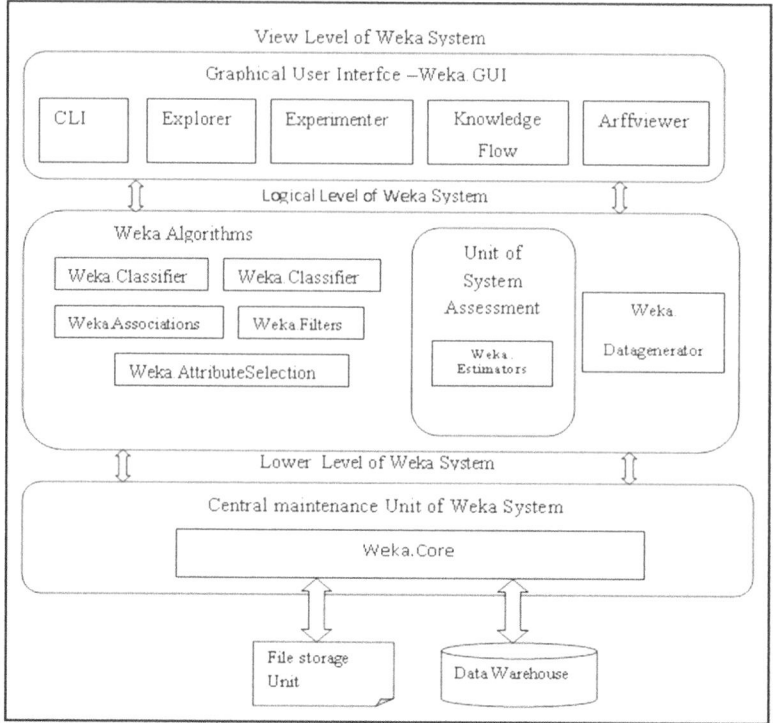

Figure 8.8 Different Components of WEKA

o WEKA.attribute selection: It is used for filtering the various attributes present in a database.
o WEKA.datagenerators: This is used to produce sample data for testing purpose.
o WEKA.estimator: It includes assessment methods that are utilized by other packages.

o WEKA.experment: This include WEKA experimenter tool, which is one of the important WEKA tool. Figure 8.8 displays different components of WEKA system.

8.4.2 Graphical User Interface of WEKA

WEKA has four components namely Simple CLI, Explorer, Experimenter, and Knowledge Flow. Figure 8.9 displays GUI components of WEKA and Figure 8.10 provides a clear view of WEKA application, i.e., Explorer, Experimenter,

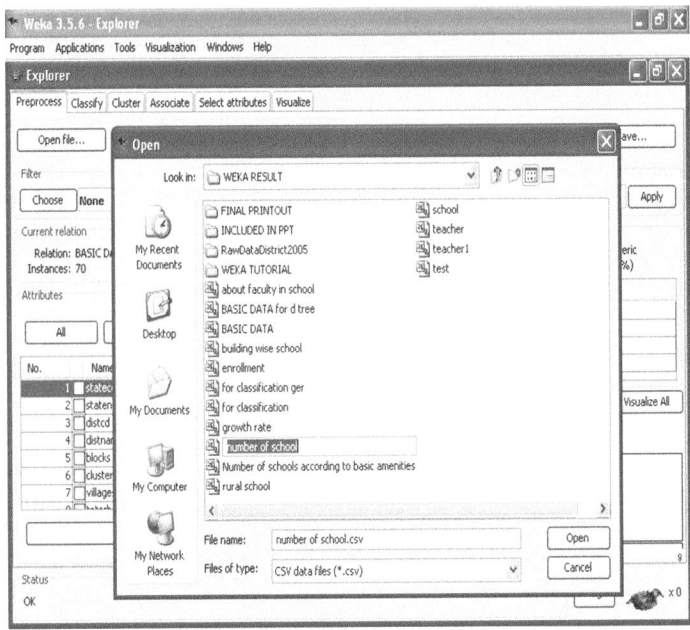

Figure 8.9 Loading of Database in WEKA Explorer

Knowledge Flow and Simple CLI [332]. With the help of Explorer, data could be mined interactively. It provides access to different algorithms for Classification, Clustering and Association. In addition to all above methods, attribute selection and data visualization could also be performed by using WEKA Explorer. The explorer first selects a filter to preprocess a data source. After selection any Data Mining algorithm could be chosen and performance analysis may be done. The Figure 8.11 shows data preprocessing tool, data filter, current relation and attribute selection tools. All selected attributes could be visualized by using data visualization. It also performs efficient handling of missing values. WEKA Experimenter provides an interface to compare different classifiers automatically. The classification accuracy of the classifier has been compared by its analyzer interface. Knowledge Flow is an interface which servers Data Mining features in very efficient manner. It uses a GUI for performing the various operations. It allows selection of any WEKA components and drags it into the working area. It also allows selection of parameters and display diagrammatic representation of all components. There is a simple command-line interface (CLI). This executes all WEKA commands from its command prompt.

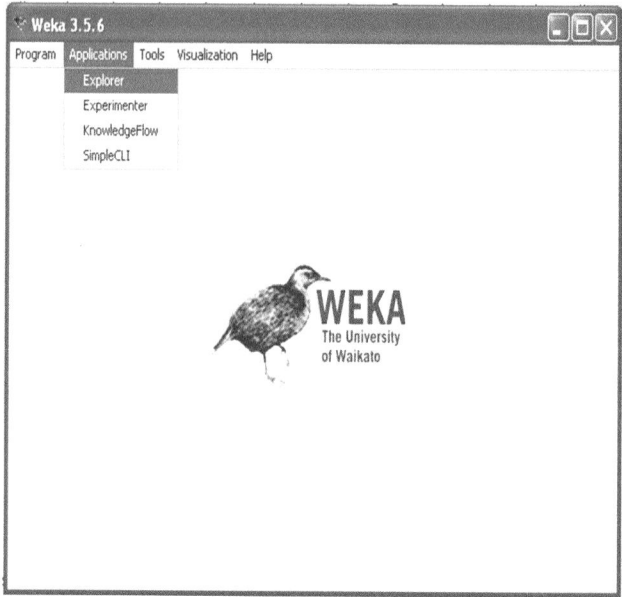

Figure 8.10 A View of WEKA GUI

8.5 Case Study: Data Mining in Department of Education

Education has significant application domain for Data Mining because it collects and generates huge data on students enrollment, courses taught, students academic record history and so on [333]. The application data has been increasing, especially due to enhanced quality, availability and popularity of the courses taught. Today many institutions also have websites where students may study online. Educational Data Mining may help identify students' academic performance and discover students' behavior regarding selection of subjects to promote quality of education, achieve better student admission and satisfaction and enhance good academic practices and policies. Data Mining algorithms are used to distinguish hidden patterns within datasets.

For instance, a Data Mining approach can distinguish students in terms of their loan liabilities to predict rules for approval of a study loan. The rule is derived from the previous records of good students who have successfully paid their loans. Likewise, various algorithms are implemented to convert the database into clusters based on similar attributes of the students and this certainly would reflect remarkable and surprising patterns. The data miners, in association with institutions personnel, then interpret the patterns discovered

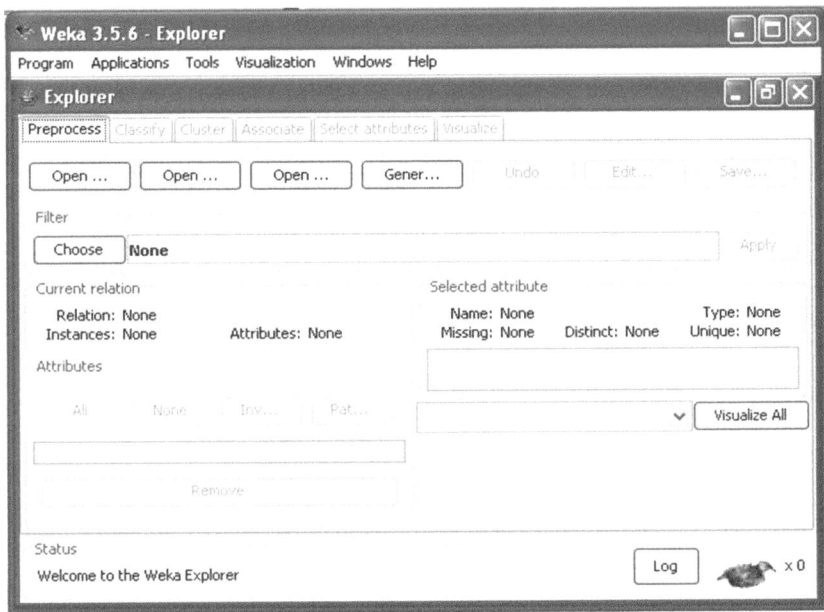

Figure 8.11 WEKA Explorer with Data Preprocessing Options

Table 8.1 Education Structure in India

Education Level	Class
Primary Education	I to V
Upper Primary Education	VI to VIII
Secondary Education	XI and XII
Higher Secondary Education	XI to XII
Higher Education	After completing XII class

by these clusters. The research work includes exhaustive study about the status of elementary education in Uttar Pradesh. The present school education structure is shown in Table 8.1 [334].

Literacy is the prime necessity for growth of any society and there were various programs conducted for 100 % literacy. "Education for All" is one such program that was targeted to achieve 100% literacy by 2010 but could not accomplish it [335]. Hence, the importance of Data Mining is obvious in preparing realistic future plan for different category of students. Present study

Table 8.2 Education Indicators in India

Indicators	Formula
Pupil/Teacher Ratio (PTR)	$\dfrac{\text{Total Enrolment in school's of Primary category}}{\text{Total Teacher in Primary School category}}$
Student Classroom Ratio (SCR)	$\dfrac{\text{Total Enrolment In Primary School}}{\text{Total Class room in Primary School}}$
Gross Enrolment Ratio (GER)	$\dfrac{\text{Total enrolment in Grades I–y}}{\text{Population of age 6–11 years}}$
Net Enrollment Ratio (NER)	$\dfrac{\text{Total enrolment in Grades I–V/6 10–11 age group}}{\text{Population of age 6–11 years}}$
Drop out Rate	$\dfrac{\text{Total enrolment in any grade in a year}}{\text{Total dropout from that grade in that year}}$

includes Data Mining for investigating the status of elementary education based on following indicators as described in Table 8.2 [335]:

The key indicators are as follows:

Total Literacy : The Present data available indicates that the total literacy is 99% in Sweden, 98.7% in Japan, 99% in USA, 99% in Norway, 99% in Canada however only 65.3% in India [336]. To achieve 100% literacy in India, a comparative study of present status of elementary education in above mentioned countries could be done and its outcome should be thoroughly examined.

Pupil Teacher Ratio: The present data available indicates that teacher to student ratio varies between 1:10, in Sweden, 1:19 in Japan, 1:14 in USA, 1:11 in Norway, 1:17 in Canada and 1:40 in India [336]. Its impact should be thoroughly examined especially for Scheduled Caste, Schedule Tribe and differently abled students because of their limited ability in teaching process. Similarly availability of learning resource in visual/audio mode will also have impact on full coverage of students.

8.5.1 Database used in the Proposed Model

The overall database is divided into following categories [335]:

Basic Data: Number of District, Number of Blocks, Number of Clusters, Number of Villages, Total Number of Schools, Total Population, 0-6

Population, Percentage Urban/SC/ST Population, Overall Literacy, Female Literacy, Sexratio, Sex Ratio 0-6, Decadal Growth Rate

School Data: Total Schools Government, Total Schools Private, Rural Schools Government, Enrolment in schools without building, Rural Schools Private, Number of Single Classroom Schools, Number of Schools with <=50 Students, Number of Single Teacher Schools, Number of Schools with PTR>100, Number of Schools with pre-primary sections, Total Classrooms.

Enrollment Data: Total Enrolment Government, Total Enrolment Private, Rural Enrolment Government, Rural Enrolment Private, Enrollment of children with disability (Girls), Enrollment of children with disability (Boys), ST Enrollment, SC Enrollment, Number of Repeaters, OBC Enrolment, GER, NER, Number of Girls, Girls Enrollment and Total Enrollment from 2001 to 2009.

Teacher Data: Regular Teachers Male, Regular Teachers Female, Teachers Private, Teachers Government School, SC Teachers Male, SC Teachers Female, ST Teachers Male, ST Teachers Female.

Results

8.5.2 Data Mining by Using Classification

Data Classification is a supervised learning and divides data for target categories or classes. The purpose of classification is to correct prediction for the target class for each data point. Binary and multilevel are the two methods of classification generally used in such investigation. In binary classification, two possible classes: "high" or "low" literacy rate may be considered. Multiclass approach has more than two targets: "poor", "average" and "high" literacy rate. In real time data analysis, Multi label data classification is more common and useful. Support Vector Machines include efficient classifier with binary and multilevel data classification. There may be two types of classification need: when data is linearly separable and when data is linearly inseparable. SVM is equipped with Kernel functions which convert non linear separable data into separable form. Three different types of Kernels, Gaussian, Polynomial and Hybrid Kernel, have been used in present study and Classification performance has been tested on the basis of Performance index Parameters.

Root Mean Square Error and Mean Absolute Error have been observed for performance analysis and it is desired to be as minimum as possible. A Multilevel data classification has been performed on the basis of Total Number of Schools, Total Number of Teachers, Gross Enrollment Ratio (GER)

Table 8.3 Overall Comparison of all Classifier in Terms of MAE and RMSE with Different Kernel Type

Classification Type	Mean Absolute Error	Root Mean Sqaure Error
SVMP E	0.4318	0.4345
SVMG E	0.4312	0.4339
SVMH E	0.4257	0.4284
WSVMP E	0.194	0.2402
WSVMG E	0.1938	0.2399
WSVMH E	0.2372	0.1915
LSSVMP E	1.7525	1.7568
LSSVMG E	0.2475	0.2759
LSSVMH E	0.1227	0.1318
WLSSVMP E	0.6055	0.7387
WLSSVMG E	0.058	0.0602
WLSSVMH E	0.033	0.037

ratio, Net Enrollment Ratio (NER) ratio, Overall literacy and Female Literacy. Algorithmic performance of classification approach is discussed in following sections.

8.5.3 Overall Performance of Different Classifiers

Performance analysis of all classifier stipulates that there is significant decrease in classification error when shifting from traditional SVM to Weighed SVM and traditional LS- SVM to weighted LS-SVM. Error could be further minimized with Hybrid Kernel. The Table 8.3 shows the overall comparison of different classifiers.

It is obvious that LS- SVM is not leading to the desired results with Polynomial Kernel. Hence, it is recommended that LS-SVM should never be used along with Polynomial Kernel. Most efficient classification has been achieved with Weighted Least Square Support Vector Machine using Hybrid Kernel, Table 8.3 shows that there is gradual reduction of classification error when various optimization criteria are applied. Figure 8.12 and 8.13 show the comparison of MAE of all classifiers with different Kernel types.

8.5.4 Data Classification for Literacy Rate

In Figure 8.14, overall literacy has been classified as below 60% and above 60%. Purple area indicates the literacy rate greater than 60% and blue area

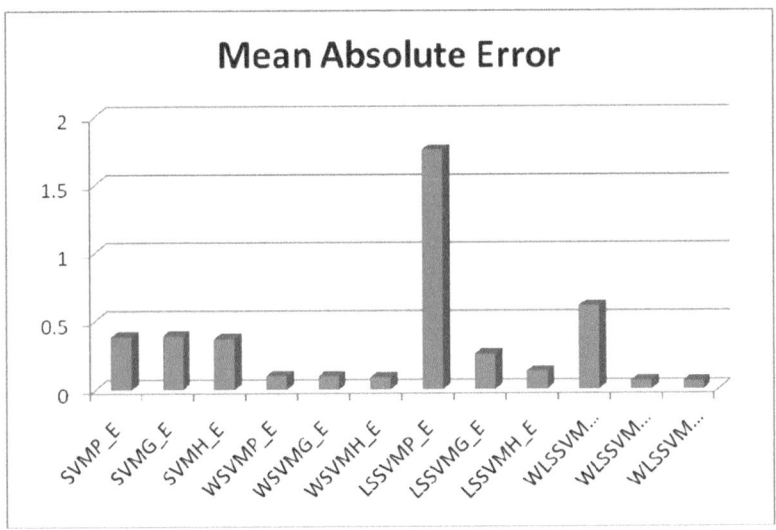

Figure 8.12 Comparison of MAE of all Classifiers with Different Kernel Types

indicates the literacy rate lees than 60%. Whereas X-axis stands for number of schools and Y-axis signifies total number of teachers. Here, it is clearly highlighted that literacy rate is high where number of school as well number of teachers are higher in number.

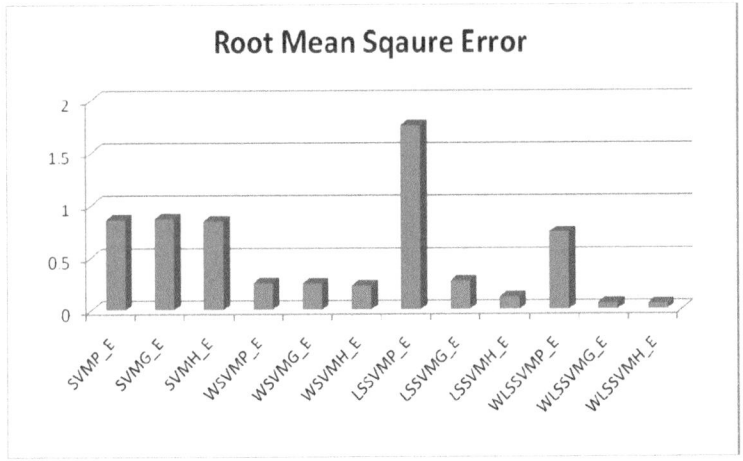

Figure 8.13 Comparison of RMSE of all Classifiers with Different Kernel Types

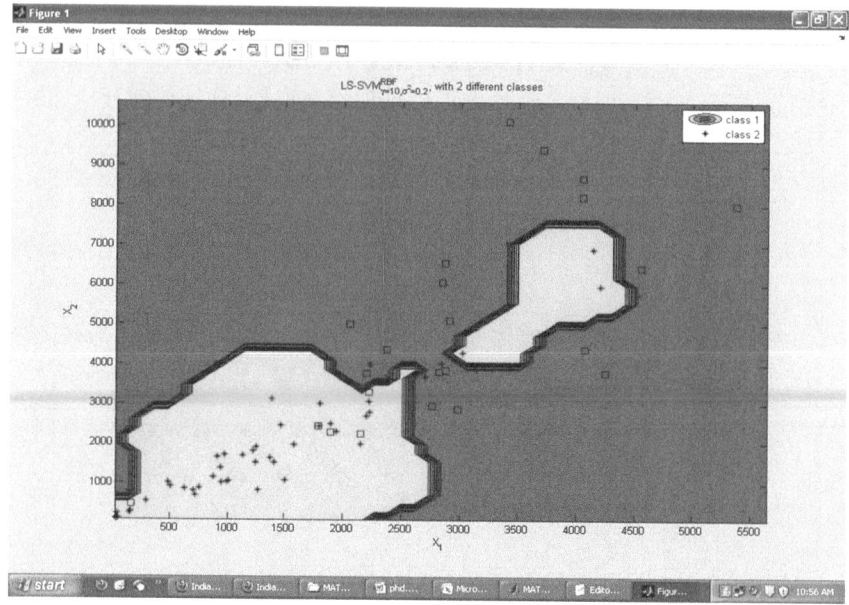

Figure 8.14 Data Classification for Literacy Rate

8.5.5 Data Classification for Decadal Growth Rate

Decadal growth has been considered as a constraint for better living standards. Lower the Decadal rate higher the index value and superior the District in terms of literacy. The study indicates that Kerala has minimum Decadal growth. Female literacy also depends upon overall literacy as indicated in the Figure 8.15. X-axis represents overall literacy rate and Y-axis represents female literacy rate. Blue region indicates decadal rate above 25% and purple region is for decadal rate below 25%.

8.5.6 Rule for Various District According to Gross Enrollment Ratio (GER) by Using Decision Tree

Decision Trees are graphical representation of data classification which shows multiple path to explore different specified conditions. It starts with a highest node where certain condition is tested and as per results different branches may be established. All conditions must be checked at the nodes moreover the leaves of the tree indicate different outcomes. The Figure 8.16 indicates a Decision Tree classification based on Gross Enrollment Ratio.

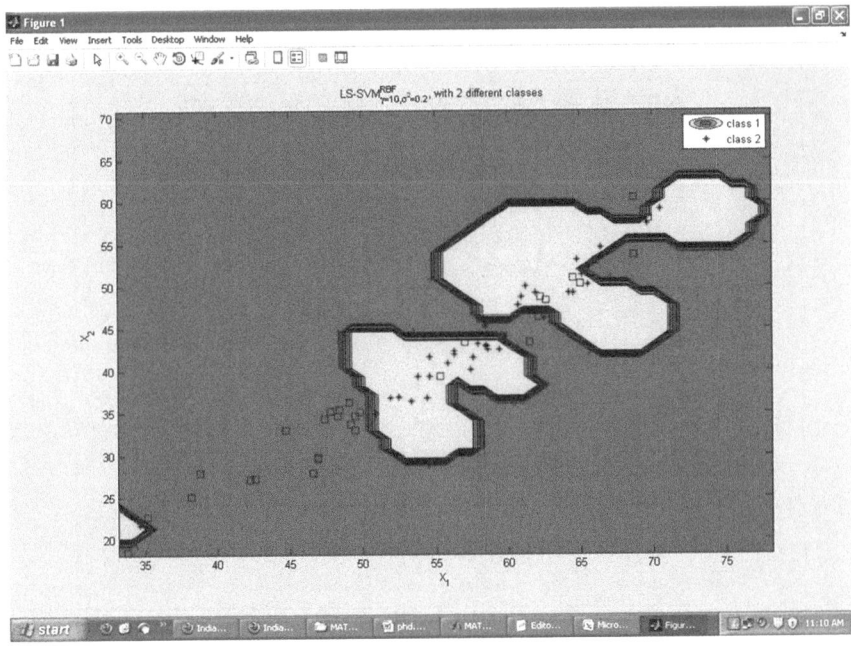

Figure 8.15 Data Classification for Decadal Growth

Here group of various Districts are classified according to their Gross Enrollment Ratio (GER). The decision rules for classification divide the groups of Districts into classes, i.e., "vary good", "good", "average", "poor" and "very poor". The Figure 8.16 indicates a Decision Tree classification based on Gross Enrollment Ratio (GER). However, Mahoba, Ambedkar Nagar, Lalitpur, Pratapgarh, Barabanki are placed in "very good" category where Gross Enrollment Ratio (GER) is between 101 to 118.99 and Lucknow, Varanasi, Meerut, Gaziabad, Allahabad, Gautam Buddha Nagar find a place in "very poor" category where Gross Enrollment Ratio (GER) is between 45 and 60.99. It appears that the above position is due to migration of students to new places for better educational facilities.

So the incidence of migration and its impact could also be highlighted here. It is also essential to trace the movement of students by using a unique identification number. The Data in Data Warehouse based on social security number will eliminate any scope for duplication leading to more reliable and dependable for strategic planning for improving the percentage education in primary sector through "Education for All" scheme.

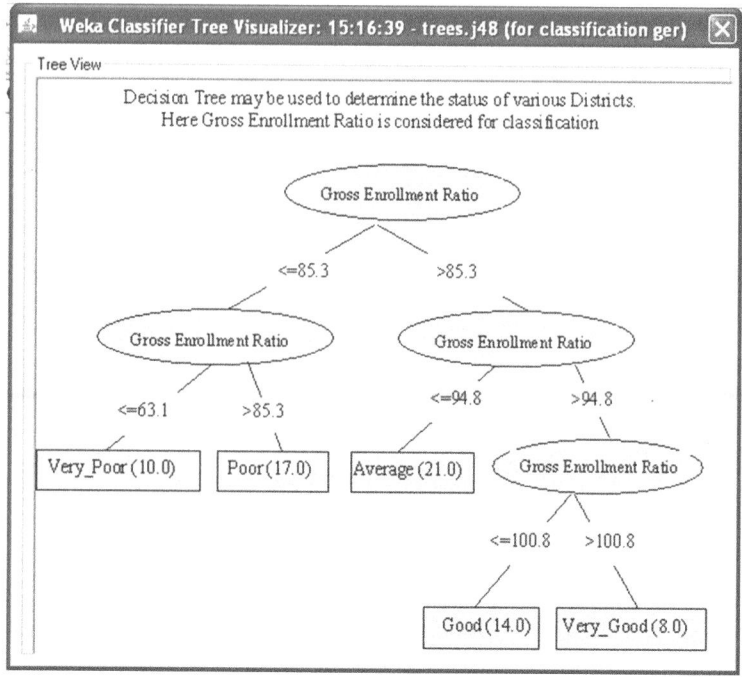

Figure 8.16 Categorization of District According to Gross Enrollment Ratio (GER) by Using Decision Tree

8.5.7 Data Mining Based on Regression

Regression is a statistical method which investigates relationships between variables. By using regression, dependences of one variable upon others may be established. Here, SVM regression has been used for function estimation and model prediction and overall Literacy Rate, Female Literacy Rate, GER and NER have been predicted. Like SVM Classification, SVM Regression can also handle data linearly separable or inseparable cases with the help of Kernel functions. The similar combination of Kernels, Gaussian, Polynomial and Hybrid have been used in present study and Regression performance has been tested on the basis of Performance index Parameters. Root Mean Square Error and Mean Absolute Error have been observed for performance analysis and it is desired to be as minimum as possible. A Regression has been performed to predict GER ratio, NER ratio, Over all literacy and Female Literacy on the basis of Total Number of Schools and Total Number of Teachers. Algorithmic performance of Regression approach is discussed in subsequent sections.

Table 8.4 Overall Comparison of all SVM Regression Techniques

Regression Type	MAE	RMSE
SVRP E	0.4318	0.4345
SVRG E	0.4312	0.4339
SVRH E	0.4257	0.4284
WSRMP E	0.194	0.2402
WSRMG E	0.1938	0.2399
WSRMH E	0.1442	0.1015
LSSVRP E	0.2991	0.2558
LSSVRG E	0.1463	0.1642
LSSVRH E	0.1015	0.1442
WLSSVRP E	0.0953	0.1055
WLSSVRG E	0.0873	0.1173
WLSSVRH E	0.0494	0.0971

8.5.8 Overall Comparison of all SVM Regression Technique

In the similar fashion of SVM Classification, performance analysis of all regression approach stipulates that there is noteworthy decrease in these errors when shifting from traditional SVR to Weighed SVR and traditional LS-SVR to weighted LS-SVR. Error could be further minimized with Hybrid Kernel. The Table 8.4 shows the overall comparison of different Regression techniques.

Here also, best results have been achieved with Weighted Least Square Support Vector Regression using Hybrid Kernel. Table 8.4 shows that there is gradual reduction of classification error when various optimization criteria are applied. Figure 8.17 and 8.18 illustrate the performance graphically.

8.5.9 Data Mining by Using Clustering

Clustering is a Data Mining approach which creates clusters of data instances within a dataset. Clusters are closed occurrence of data items under the consideration of certain parameters. These clusters further represent similar groups. In this study, raw data of education for Uttar Pradesh have been taken. The database has 70 instances, which represent all 70 Districts of Uttar Pradesh. In the proposed approach various Districts may be clustered according to their similarity. These groups of Districts as clusters may be governed collectively under one policy.

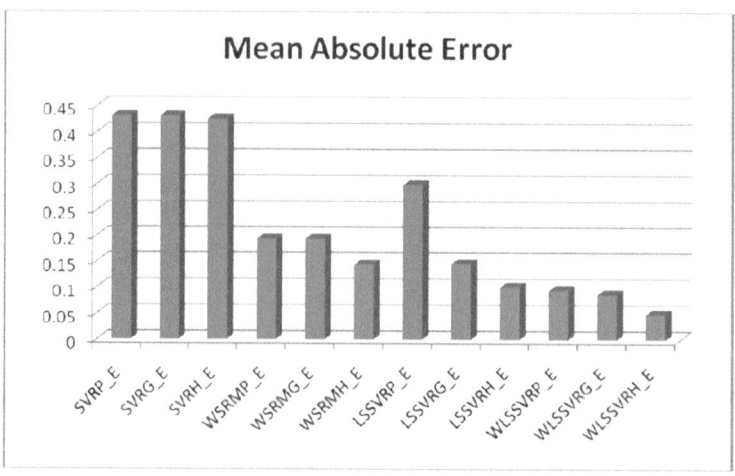

Figure 8.17 Overall Performace of Different SVM Classification in Terms of Mean Absolute Error

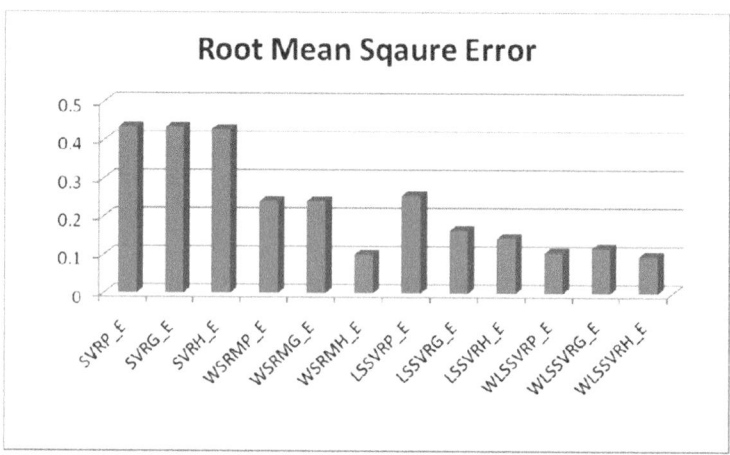

Figure 8.18 Overall Performace of Different SVM Classification in Terms of Root Mean Square Error

8.5.10 Performance Analysis of Fuzzy Clustering

The performance of Fuzzy C Means Clustering has been evaluated by using proposed Performance Index Parameters. There are 7 performance indicators namely Partition Coefficient, Partition Index, Classification Entropy, Separation Index, Xie and Beni's Index, Dunn's Index and Alternative Dunn Index.

These Performance Index Parameters prove validity of each cluster. Clustering performance also depends upon correct choice of number of clusters.

By using above Performance Index Parameters, number of clusters can be easily estimated. For efficient clustering, lesser Partition index and XB Index is required whereas high Partition Coefficient and Classification Entropy is desired. The Table 4.11 shows that Partition index and Xie and Beni's Index is less for cluster number 8 and 9 so here, it is suggested that For "Education For All" dataset where 72 Districts have been considered for analysis, they may be grouped in terms of 8 or 9 clusters on the basis of Literacy Rate.

8.5.11 Performance Analysis of Optimized Weighted Fuzzy Clustering

Additionally, a novel Fuzzy C Means approach has been developed with modified distance factor, weighted cluster centre and exponent. Here, it has been observed that for calculation of degree of membership "Harmean" distance shows better results than "Mean" distance. Similarly, the calculation of cluster centre has also been modified by using a weight function. For estimating the centroids, the weighted clusters are helpful in suppressing the undesirable effect of outliers. A novel way of estimation of weighting exponent m has been proposed as fuzziness factor and it should always be greater than one.

All these modifications showed better results as compared to a conventional Fuzzy C Means and the new modified technique has been termed "Optimized Weighted Fuzzy C Means (OWFCM)". As per Table 8.5, it is noted that Partition index and Xie and Beni's Index is further minimized for same 8 and 9 clusters and Performance Coefficient is comparatively increased to indicate better performance.

8.5.12 Clusters Based on Number of Enrollment from Class 1 to 6

As per number of enrollment from class 1 to 6 in primary education, the different Districts of Uttar Pradesh are grouped into following clusters as mentioned in Figure 8.19.

8.5.13 Overall Cluster Based on Overall Literacy

According to the literacy rate, the different Districts of Uttar Pradesh are grouped into following clusters as mentioned in Figure 8.20.

Table 8.5 Identification of Optimum Number of Clusters in Fuzzy C Means Clustering

Performance Index Parameter	Number of Clusters					
	2	3	4	5	6	7
PC	0.8189	0.7961	0.7397	0.7406	0.7275	0.7344
CE	0.3013	0.3708	0.5042	0.5179	0.5635	0.5632
SC	3.82E-007	2.18E-007	2.05E-007	1.43E-007	1.40E-007	9.38E-008
S	5.54E-009	5.03E-009	4.80E-009	3.34E-009	3.28E-009	2.20E-009
XB	5.5255	6.1135	6.7960	4.0648	3.3302	3.8866
DI	0.0520	0.0520	0.0639	0.0291	0.0991	0.1862
ADI	1.16E-004	1.36E-004	1.50E-004	1.81E-004	2.62E-005	5.17E-005

Performance Index Parameter	Number of Clusters						
	8	9	10	11	12	13	14
PC	0.6196	0.6207	0.6445	0.6579	0.6620	0.6628	0.6576
CE	0.7942	0.8189	0.8033	0.7S62	0.7689	0.7736	0.857
SC	2.08E-007	1.18E-007	9.01E-008	1.81E-007	024E-008	9.56E-008	8.28E-008
S	5.0OE-009	2.99E-009	2.26E-009	3.75E-009	1.75E-009	2.02E-009	1.91E-009
XB	2.8322	2.1274	7.9173	2.5521	2.9583	2.0706	1.8562
DI	0.0139	0.0638	0.1726	0.0521	0.0749	0.0461	0.1264
ADI	9.S1E-006	3.53E-005	6.47E-006	1.09E-006	1 49E-005	30E-005	4.05E-007

Table 8.6 Identification of Optimum Number of Clusters in Optimized Weighted Fuzzy C Means Clustering

Performance Index Parameters	Number of Clusters					
	2	3	4	5	6	7
PC	0.8664	0.7745	0.7857	0.761	0.7351	0.7463
CE	0.2302	0.4062	0.417	0.4728	0.5438	0.537
SC	2.43E-07	1.85E-07	1.20E-07	1.09E-07	1.07E-07	7.94E-08
S	3.52E-09	4.54E-09	2.50E-09	2.62E-09	2.42E-09	1.93E-09
XB	9.2331	5.3685	4.3594	8.0792	2.0863	9.3651
DI	0.0802	0.0117	0.0335	0.0726	0.1563	0.1204
ADI	4.30E-05	1.56E-04	2.40E-04	1.48E-04	1.87E-05	1.36E-05

Performance Index Parameters	Number of Clusters					
	8	9	10	n	12	13	14
PC	0.7549	0.7606	0.7168	0.7256	0.6913	0.6976	0.7237
CE	0.5335	0.5381	0.63	0.6198	0.7081	0.6892	0.6504
SC	7.40E-08	7.28E-0S	6.63E-08	6.40E-08	7.13E-08	7.52E-08	6.40E-08
S	1.73E-09	1.72E-09	1.63E-09	1.50E-09	1.71E-09	1.69E-09	1.44E-09
XB	1.7913	2.3837	1.706	1.8304	1.4184	2.6996	5.3323
DI	0.1611	0.1705	0,1598	0.1705	0,1812	0,1295	02447
ADI	2.43E-05	9.62E-05	7.38E-06	1.06E-04	5.52E-06	1.89E-05	9.27E-06

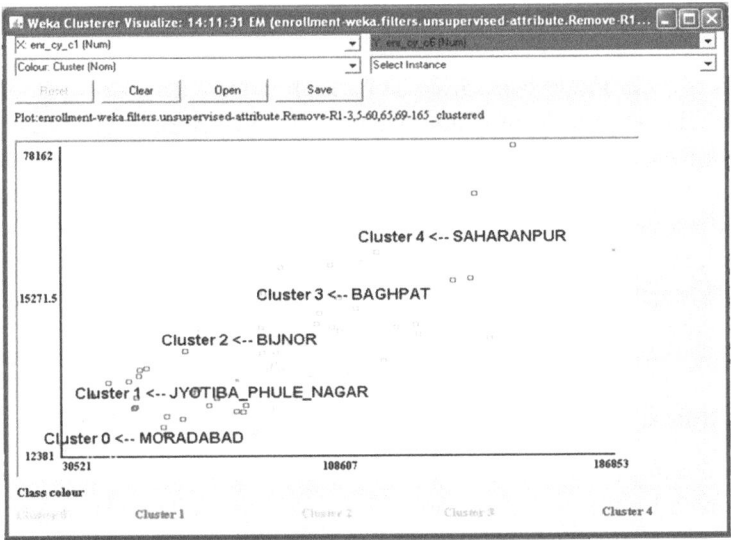

Figure 8.19 Clusters Based on Number of Enrollment from Class 1 to 6

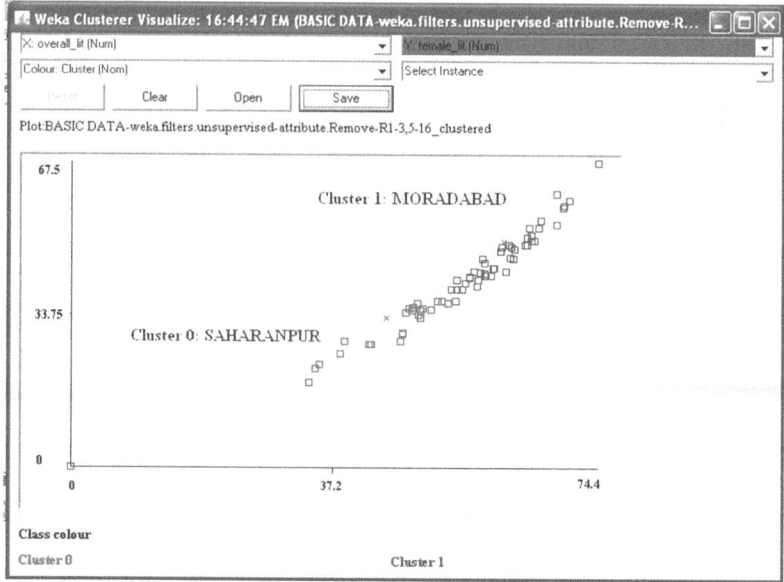

Figure 8.20 Cluster Based on Overall Literacy

8.5.14 Cluster Based on Basic Amenities Present in School

According to the basic amenities present in primary schools, the different Districts of Uttar Pradesh are grouped into following clusters as mentioned in Figure 8.21 The Data Mining approach based on clustering clearly indicates significant variations between clusters of Districts from another cluster. However, the cluster approach could be sharpened when data for each District-rural, urban; category wise-general, OBC, SC, ST, visually impaired, hearing impaired, mentally retarded are classified on account of social security number to have qualitative approach to entire planning and accomplishment of "Education for All" program.

8.6   Case Study: Data Mining in Department of Health

In the present scenario Data Mining is becoming choice of every domain and adopted by many organizations intensively as well as extensively. The health domain also consists patient records. The data may include demographic Information, disease related information, status of health care measures etc. These records are then converted into standard datasets for serving

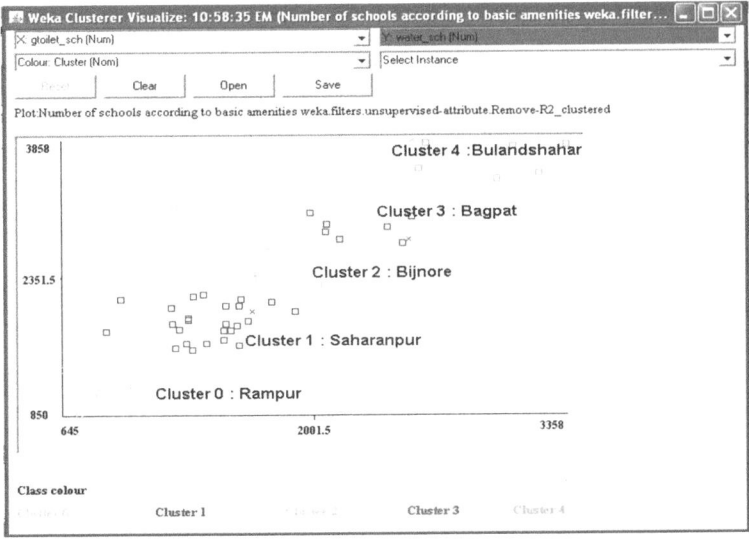

Figure 8.21 Clusters Based on Basic Amenities Present in Schools

various heath information systems. Transaction data generated by health information systems are huge, complex and difficult to process by traditional methods. Data Mining offers well defined set of approaches and expertise to explore valuable information from voluminous data for decision making [337].

As per Vision 2020, ideal health care system should have four important characteristics [338]. The prime necessity is accessibility of health services to everyone without having any burden. Second priority is fair establishment of cost effective health care services. Third important consideration is regarding quality of health care products. Lastly it is also raised the need of special attention to weaker sections such a senior citizens, children, women and disabled. Better planning of health services needs proper understating of health indicators and their distribution throughout the country. Uneven distribution of health indicators requires special attention in certain sectors. The common health indicators are life expectancy, infant mortality rate, birth rate, no. of hospitals, no. of heath care experts and disease related information. Data Mining could help in analyzing, predicting and planning health care measures based on above indicators. Hence, the end objective of this case study is to investigate Data Mining in healthcare through Classification, Clustering and Regression techniques.

Table 8.7 Latest Status of Polio

Total Cases Wild Polio Virus	2009	Jan 01-13 Aug 2009*	Jan 01-13 Aug 2010*
Globally	1606	817	608
Pakistan	80	31	36
Afghanistan	38	14	13
India	741	184	27
Nigeria	388	372	6
No of countries with virus.	23	19	15

8.6.1 Introduction about "Pulse Polio Immunization" Program

In 1988 Pulse Polio Program was initiated by World Health Assembly, to eliminate Polio completely from the whole world. India is Polio-endemic country which has more than 50% of Polio cases reported globally. In 1994 Pulse Polio immunization program was commenced by government of India to exterminate Poliomyelitis. It has annual vaccination plan intended for every children under age five. In this program every child has given an oral dose to restrain the wild, disease-causing Polio virus [316] [339].

The basic plan of Pulse Polio Immunization is as follows:

- Three does of oral Polio vaccines must be given to every child below 1 year
- Additional 2 oral doses must be given to every child below 5 years on National Immunization Days.
- Identification of all means caused transmission of Poliovirus
- Extensive Polio camps to conduct house-to-house campaigns.

There is substantial improvement in status of Polio after a Pulse Polio Immunization program. In the year of 1998 worldwide reported Polio cases were 350,000 but now it was recorded only 1604 in 2009. As per latest studies of 2010, Total no. of Polio infected countries are 23 but four countries remain more Polio endemic. Earlier in 1998, it was 125. So there is noticeable reduction in Polio prone areas. The current status of Polio worldwide is shown in Figure 8.22.

The countries still facing Polio are India, Pakistan, Nigeria and Afghanistan. Table 8.7 shows latest status of Polio cases.

There is considerable reduction in Polio cases from 2009 to 2010 but still the 100% Polio free goal has not been achieved. The Figure 8.22 also indicates different wild virus type and corresponding endemic countries [339] [340]. The Figure indicates that India has highest number of Polio cases which has to be really considered as serious problem for the country. The Pulse Polio

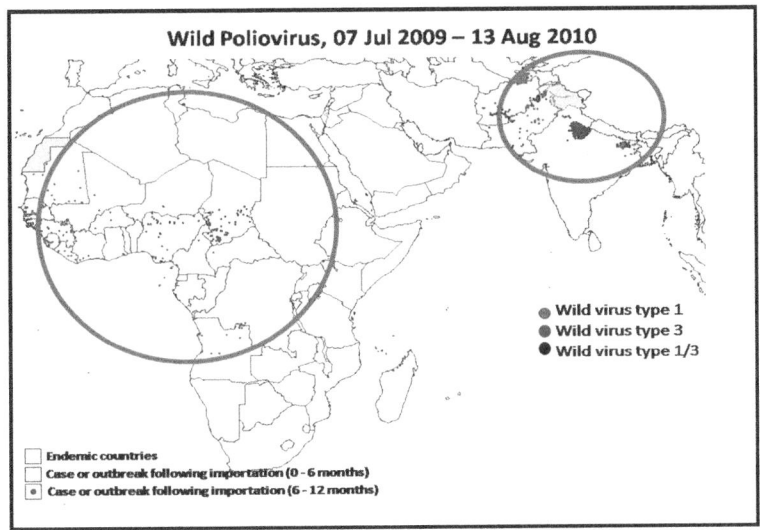

Figure 8.22 Worldwide Polio Status

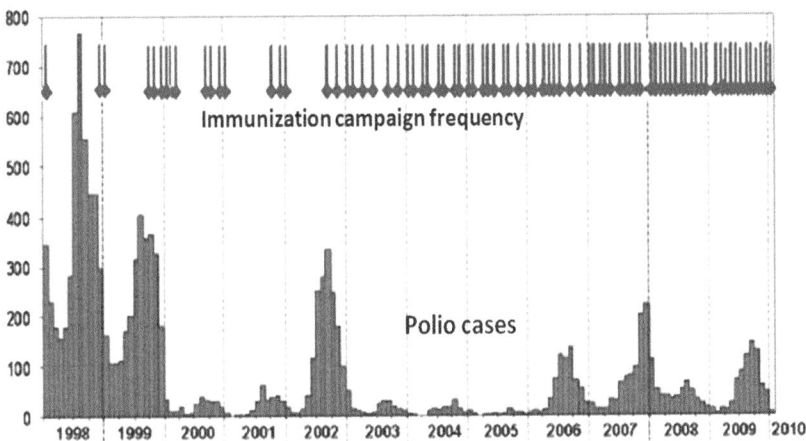

Figure 8.23 Number of Polio Cases Reported in India

Program proved to be doing well, and the Poliomyelitis cases in India has dropped off dramatically as indicated in the Figure 8.23.

The Figure 8.24 indicates Polio cases of India since last 6 years.
The Figure indicates that in the 2000, 2001, 2003, 2004 and 2005 was the huge success for Pulse Polio Immunization Program. But in the year of 2006, 2007,

2008 and 2009 it started increasing that may due to the developed resistance against the Polio doses. This is because Polio is one of the communicable disease and spread among children very quickly. There was substantial growth in Polio during 2006, 2007, 2008 and 2009. The Figure indicates that many states are Polio free now but Bihar and Uttar Pradesh have extremely large number of Polio cases and it should be the prime focus for Polio eradication initiatives [339]. Table 8.8 indicates that there were total 10 cases within 7 districts and 7 blocks. Most of these districts are situated at western Uttar Pradesh and geographically closed. In Figure 8.25 various districts of Uttar Pradesh have been examined on the basis of reported Polio cases for the year of 2010.

Figure 8.26 indicates that Polio strain may also be correlated with the regional communities, Gender and Rural Urban location. Here, it is clear that Polio cases are more in Muslim community. The share of female is also on higher side. It is also obvious that due to living standards and hygiene it is more prone to rural areas. The Polio related databases have been collected

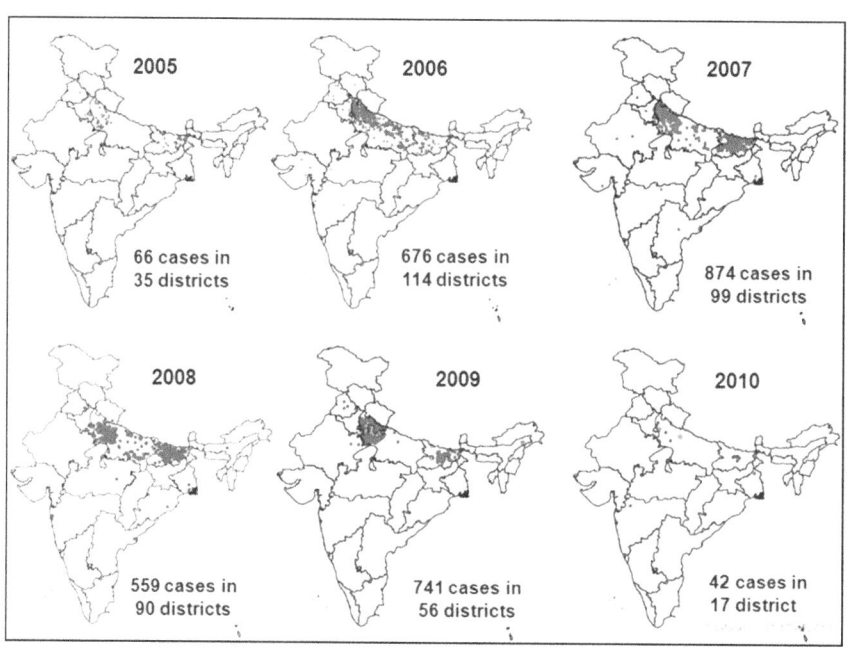

Figure 8.24 Polio Cases Reported in Last Six Years

Table 8.8 Status of Polio in Uttar Pradesh

District	P1	P3	P1+P3	Total Cases
BADAUN	0	1	0	1
BAGHPAT	0	1	0	1
ETAH	0	1	0	1
FEROZABAD	0	1	0	1
GHAZIABAD	0	2	0	2
MATHURA	0	2	0	2
MUZAFFARNAGAR	0	2	0	2
Total: UP	0	10	0	10

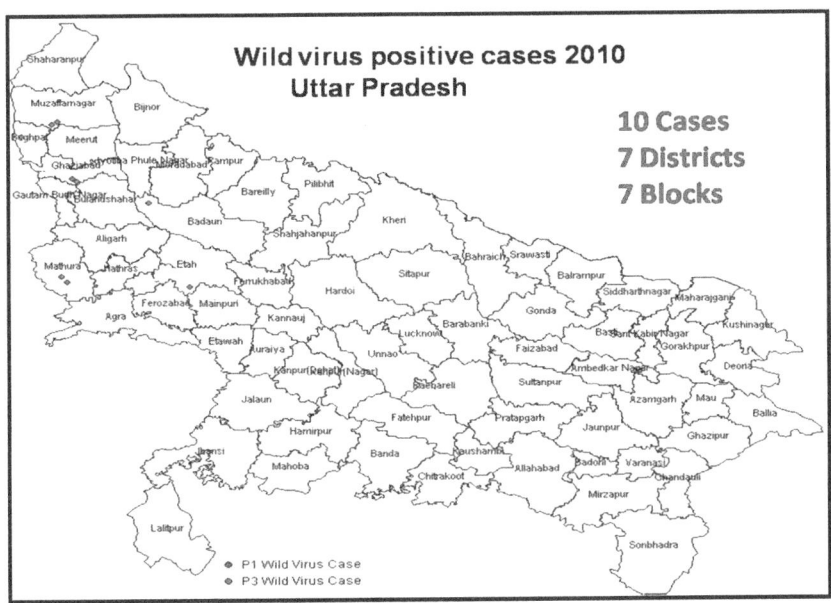

Figure 8.25 Polio Cases in Uttar Pradesh for the Year of 2010

from official National Polio Surveillances Project, a Government of India, WHO collaboration project [316] [339].

8.6.2 Objective of Study

The status of "Pulse Polio Immunization" program under Data Mining perspective has been studied. Although Polio has been eradicated from many states of India but Bihar and Uttar Pradesh are still affected. The results after

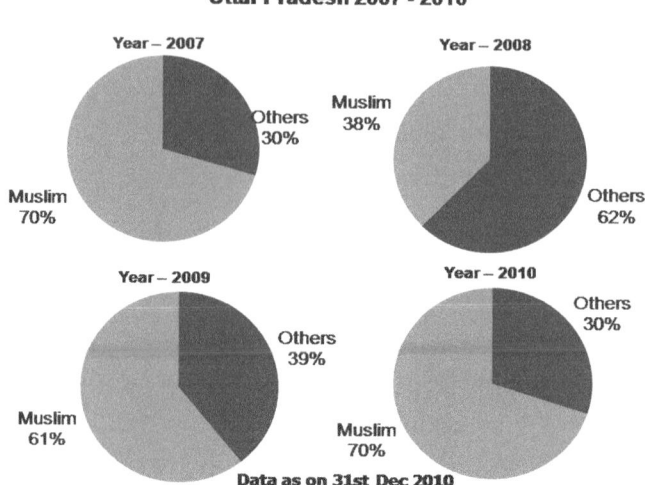

Figure 8.26 Polio Cases in Uttar Pradesh for the Year of 2007–2010

starting of "Pulse Polio Immunization" Program is undoubtly encouraging but some hurdles are still creating difficulties towards the 100% success of the whole campaign. The study exposed some social, religious and administrative constraints which are responsible for poor utilization of the services provided by WHO and NPSU. Here, it is aimed to show interesting correlation between urban, rural area, gender, religion temperature and categories of Polio along with its probability of eradication. Here, a "Pulse Polio Immunization" dataset of Uttar Pradesh have been considered to examine the functioning of FCM clustering, SVM Classification and Regression. The interesting patterns between Polio categories, region, religion, gender and seasons have been identified and discussed.

8.6.3 Database Used in the Proposed Model

The Polio related databases have been collected from official National Polio Surveillances Project, a Government of India, WHO collaboration project. The Data attributes are No. of Polio cases reported, categories of Polio, gender, area, religion, district, year wise Polio cases State, total no. of Polio cases, Confirmed Polio cases, Acute Flaccid Paralysis (AFP) Rates, Non-Polio AFP Rate, Confirmed Polio etc.[316] [339].

```
Setting Rural / Urban  = R: P3 (527.0/25.0)
Religion = M: P3 (58.0/3.0)
Hot = N AND
Age (in months) > 15: P3 (6.0)

Sex = M AND
Hot = Y AND
Age (in months) <= 16: P3 (4.0)

Sex = F: P1 (5.0/2.0)
Age (in months) > 12: P1 (4.0/1.0)
: P3 (2.0)

Number of Rules :     7
```

Figure 8.27 Classification Rule for Polio According Rural and Urban Population

8.6.4 Rules for the Different Categories of Polio Using Decision Tree

The decision tree indicates P_3 category of Polio is very high when the age is less than 54 months and P_1 category of Polio is high when age is greater than 54 months. For P_3 category of Polio, if age group is 9–14 months, most of the Polio cases belong to male children and if age group is 12–23 months, more Polio cases are recorded in the female children. For P_1 category, male children cases are more than females. Figure 8.28 indicates classification Rule for Polio categories according Rural and Urban Population.

The probability of P_3 Polio category is more in urban male children while Polio category P_1 is more common with rural females. P_3 Polio cases are more prominent if the child is Muslim, Male and belongs to high temperature areas. P_1 Polio is more common with Hindu, Female children where the weather is not very hot. Rules for various Polio categories according to regional characteristics are shown in Figure 4.38.

8.6.5 Classifications results with Pulse Polio Immunization Dataset

The Pulse Polio Immunization dataset of Uttar Pradesh with different Districts have been taken for Classification purpose. This data is slightly different from "Education for All" dataset in terms of its attribute characteristics. This dataset has more nominal values and less numerical value whereas in "Education for All" dataset mostly all attributes are numerical in nature. Due to this

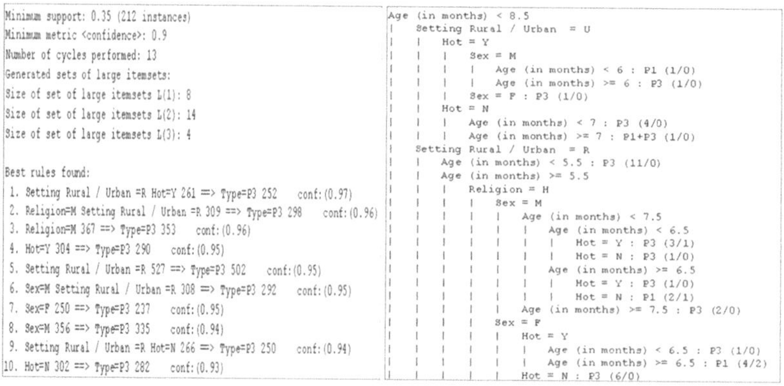

```
Minimum support: 0.35 (212 instances)          | Age (in months) < 8.5
Minimum metric <confidence>: 0.9               |   Setting Rural / Urban  = U
Number of cycles performed: 13                 |   |   Hot = Y
Generated sets of large itemsets:              |   |   |   Sex = M
Size of set of large itemsets L(1): 8          |   |   |   |   Age (in months) < 6 : P1 (1/0)
                                               |   |   |   |   Age (in months) >= 6 : P3 (1/0)
Size of set of large itemsets L(2): 14         |   |   |   Sex = F : P3 (1/0)
Size of set of large itemsets L(3): 4          |   |   Hot = N
                                               |   |   |   Age (in months) < 7 : P3 (4/0)
Best rules found:                              |   |   |   Age (in months) >= 7 : P1+P3 (1/0)
                                               |   Setting Rural / Urban  = R
1. Setting Rural / Urban =R Hot=Y 261 ==> Type=P3 252   conf:(0.97)  |   Age (in months) < 5.5 : P3 (11/0)
2. Religion=M Setting Rural / Urban =R 309 ==> Type=P3 298   conf:(0.96)  |   Age (in months) >= 5.5
3. Religion=M 367 ==> Type=P3 353   conf:(0.96)  |   |   Religion = M
4. Hot=Y 304 ==> Type=P3 290   conf:(0.95)     |   |   |   Sex = M
5. Setting Rural / Urban =R 527 ==> Type=P3 502   conf:(0.95)  |   |   |   Age (in months) < 7.5
6. Sex=M Setting Rural / Urban =R 308 ==> Type=P3 292   conf:(0.95)  |   |   |   |   Age (in months) < 6.5
7. Sex=F 250 ==> Type=P3 237   conf:(0.95)     |   |   |   |   |   Hot = Y : P3 (3/1)
8. Sex=M 356 ==> Type=P3 335   conf:(0.94)     |   |   |   |   |   Hot = N : P3 (1/0)
9. Setting Rural / Urban =R Hot=N 266 ==> Type=P3 250   conf:(0.94)  |   |   |   |   Age (in months) >= 6.5
10. Hot=N 302 ==> Type=P3 282   conf:(0.93)    |   |   |   |   |   Hot = Y : P3 (1/0)
                                               |   |   |   |   |   Hot = N : P1 (2/1)
                                               |   |   |   Age (in months) >= 7.5 : P3 (2/0)
                                               |   |   |   Sex = F
                                               |   |   |   |   Hot = Y
                                               |   |   |   |   |   Age (in months) < 6.5 : P3 (1/0)
                                               |   |   |   |   |   Age (in months) >= 6.5 : P1 (4/2)
                                               |   |   |   |   Hot = N : P3 (6/0)
```

Figure 8.28 Rule for Different Categories of Polio According to Regional Characteristics

Table 8.9 MAE and RMSE of Different SVM Classification with Hybrid Kernel

Classification Type	MAE	RMSE
Conventional SVM Classification with Hybrid Kernel	0.2166	0.1419
Weighted SVM Classification with Hybrid Kernel	0.0711	0.0485

diverse data attribute characteristics, their classification performance is also noted differently. But in this case also weight function and Hybrid Kernel are significantly helpful to improve the performance. An experiment has been performed by using conventional and weighted SVM with Hybrid Kernel and the usefulness of weight function and Hybrid Kernel have been verified. Figure 8.29 shows performance of SVM and WSVM with Hybrid Kernel.

8.6.6 Clusters Based on Polio Incidents

A "Pulse Polio Immunization" dataset of Uttar Pradesh Districts have been taken. Clustering has been performed using WEKA tool to explore the Polio incidents in Uttar Pradesh and some rules may be correlated based on the association among child age, community, location, gender and categories of Polio. The Table 8.10 displays highly Polio prone Districts as per reported Polio cases. A Clustering approach has been performed with a national level Polio dataset and 6 clusters have been achieved as on the basis of AFP rates and Non Polio AFP rates. This approach illustrates that Polio eradication policies should be formed on the basis of similarities of the states and the same groups should be considered at the time of allocating financial and human resources. Table 8.11 shows the results of clustering on national level Polio dataset.

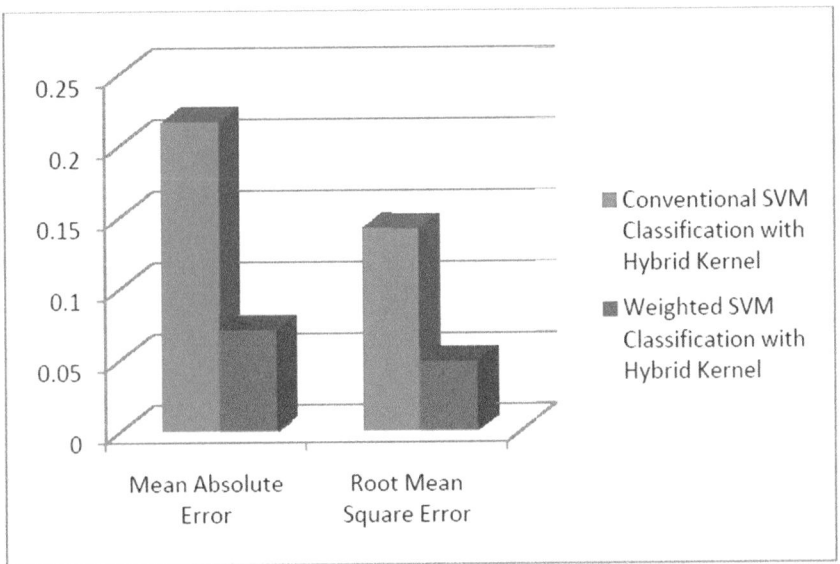

Figure 8.29 Rule for Different Categories of Polio According to Regional Characteristics

Table 8.10 District Having High Polio Cases

S.No.	Name of Disticts	Reported Polio cases
1	Ghaziabad	75
2	Moradabad	74
3	JyotibaphuleNagar	51
4	Bulandshahar	43
5	Badaun	52
6	Bareilly	36
7	Muzaffarnagar	33

A Clustering approach has been performed with Polio dataset of whole countries for years 2006 to 2009 and different clusters were achieved as per similarity of age group, gender, community, Categories of Polio and location such as rural or urban as indicated in Figure 8.30.

Another observation as shown in Figure 8.31 indicates that seasonal variation also affects the intensity of Polio. First four months of the year is showing poor polio occurrence rate whereas summer and rainy season during July to September are having high Polio occurrence rate.

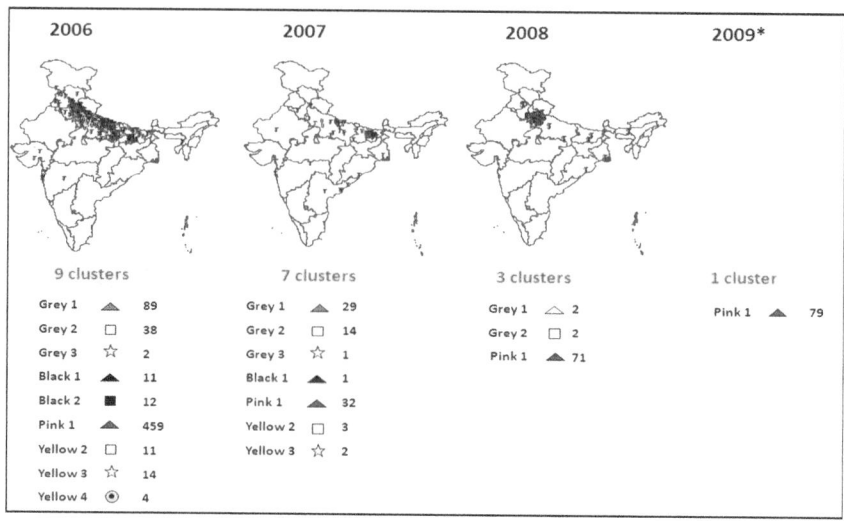

Figure 8.30 District wise Polio Cases

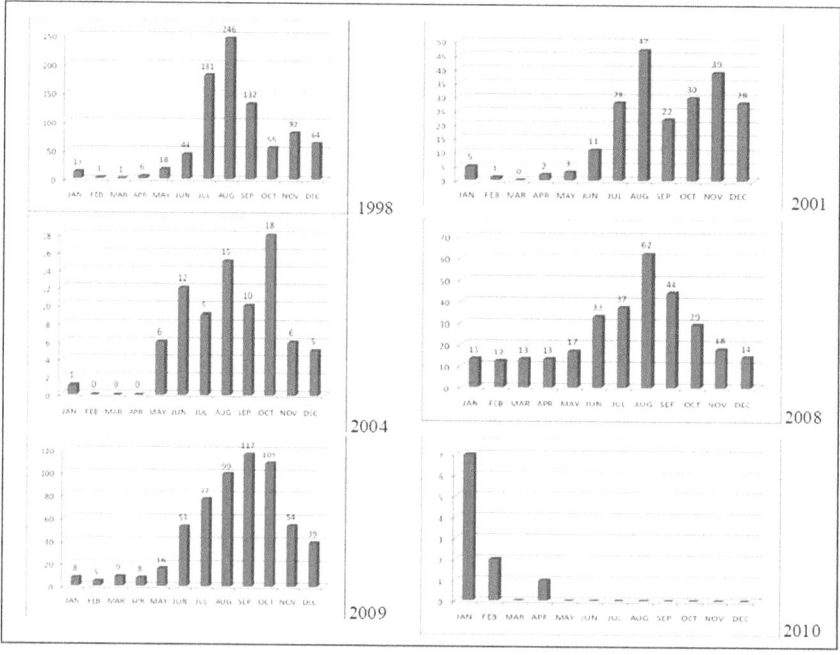

Figure 8.31 Month wise Polio Cases

Table 8.11 Clustering of National Level Polio Cases

Cluster Name	Cluster Centre	States
Cluster 0	West Bengal	West Bengal, Himachal Pradesh, Maharashtra, Haryana, Uttrakhand, Madhya Pradesh
Cluster 1	Lakshdweep	Lakshdweep, Rajasthan, Tripura, Delhi
Cluster 2	Uttar Pradesh	Uttar Pradesh, Bihar
Cluster 3	Mizoram	Mizoram, Manipur, Andaman & Nicobar, Meghalaya, Nagaland, Arunachal Pradesh, Sikkim, Chandigarh, Daman & Diu, Goa, D & N Haveli, Pondicherry
Cluster 4	Jharkhand	Jharkhand, Orrisa
Cluster 5	Andhra Pradesh	Assam, Tamilnadu, Jammu & Kashmir, Chhattisgarh, Andhra Pradesh, Kerala, Gujrat, Punjab, Karnataka

8.7 Conclusion

The present investigation deals with two case studies, i.e., "Education for All" and "Pulse Polio Immunization" to ascertain the contribution of Data Mining in E Governance. Data sets from these two projects have been selected. Clustering, Classification and Regression techniques have been used to determine similarities, dissimilarities in all 72 Districts of Uttar Pradesh. The differentiation based on cast, region, religion, degree of education, financial status or any other consideration for failure of these programs have been identified and further actions have been proposed. The results of present investigation have been designed to serve the interest of two different domains:

- An efficient GUI based MATLAB Tool has been developed, which includes Clustering, Classification and Regression techniques with several innovative optimization parameters. This tool can be utilized by Data Miners for other similar tasks and its algorithmic performance may be tested. If need arises, the tool may be further extended to solve other problems.
- The Data Mining results can also be interpreted to suit E Governance implementation needs and could also be used by the policy makers as a futuristic strategic management tool.

The tool thus developed, includes an implementation of Classification and Regression of traditional Support Vector Machine and Least Square Support Vector Machine. A traditional Fuzzy C Means Clustering has also been developed. Additionally improved SVM Classification, Regression and Fuzzy C Means Clustering have been developed to analyze the algorithmic performance and outcomes of "Education for All" and "Pulse Polio Immunization" programs.

The performance of classification for binary Support Vector Machine (SVM) and all possible Multi label SVM have been evaluated and better classification parameters have been recognized using newly proposed weighted SVM and Hybrid kernel approach. The Performance Index Parameter (PIP) has been developed to describe the efficacy of the model for a set of observations. The Performance Index Parameter (PIP) summarizes the inconsistency between recorded values and the anticipated values. Performance analysis of all classifier stipulates that there is significant decrease in classification error when shifting from traditional SVM to Weighed SVM and traditional LS-SVM to weighted LS-SVM. Error could be further minimized when using Hybrid Kernel. Most efficient classification has been achieved with Weighted Least Square Support Vector Machine using Hybrid Kernel.

Classification has been performed on reported Gross Enrollment Ratio (GER), Net Enrollment Ratio (NER), Female Literacy Rate and Over All Literacy Rate. Regression has been employed to forecast the decadal growth rate on the basis of overall literacy. Here, classification clearly establishes decision making rules to classify Districts in various categories. It is also noted that some Districts, Classification has indicated unexpected results. This could be possible due to migration of residents form one place to another. Interestingly, it is also noted that outliers may play a leading role in establishing unusual trends. Some atypical categorizations of District have been obtained due to 'citizen migration'. This needs to be considered while implementing developmental projects like "Education for All" and "Pulse Polio Immunization" for better results.

In the similar fashion of SVM Classification, Regression has been performed by using Support Vector Regression (SVR), Least Square SVR and Weighted Least Square SVR using conventional as well as Hybrid Kernel. Performance analysis of all regression approach stipulates that there is significant decrease in both Mean Absolute Error and Root Mean Square Error while using Weighted Least Square SVR in place of conventional SVR. Error could be further minimized when using Hybrid Kernel. The traditional Fuzzy C Means Clustering has been performed and a novel Fuzzy C Means approach has been developed with modified distance factor, weighted cluster centre and exponent. Here, it is observed that for calculation of degree of membership "Harmean" distance shows better results than "Mean" distance. Similarly, the calculation of cluster centre has also been modified by using a weight function. For estimating the centroids, the weighted clusters are helpful in suppressing the undesirable effect of outliers. A novel way of estimation of weighting exponent m has been proposed and it should always be greater than one. All these modifications showed better results as compared to a conventional Fuzzy C Means and the new modified technique has been termed "Optimized Weighted Fuzzy C Means (OWFCM)".

A combination of Fuzzy C Means Clustering based on Weighted Support Vector Machine with Hybrid Kernel has also been developed and its performance capability has been established. This combined approach results in smallest value of Mean Absolute Error and Root Mean Square Error as compared to approaches as mentioned above.

Clustering has been implemented to figure out Districts with similar primary education patterns or similar health conditions which may be governed under one policy. The cluster approach could prove more efficient and sharp in addressing both health and education related challenges if data for each District- rural, urban, category wise-general, OBC, SC, ST, visually impaired, hearing impaired, mentally retarded are classified with the help of unique identification number for qualitative approach to wholesome development of projects in consideration.

9

Data Warehousing and Data Mining Applications in Various Governmental Departments

9.1 Introduction

Different departments of Central and State Governments may adopt Data Warehousing and Data Mining techniques to witness the priorities and challenges of the new age. They have widespread potential purposes in diverse Central Government sectors, i.e., Health, Rural Development, Banking, Agriculture and Energy and also in the State Government activities. Data Warehousing and Data Mining applications in different departments include quicker information retrieval, fraud detection and preparing future policies based on previous experiences [341]. The major application areas of Data Warehousing and Data Mining in different government departments are as follows:

9.2 National Data Warehouse

It is essential to have National level Data Warehouse which has accurate and reliable datasets for various applications in Data Mining techniques. In developing countries like India which is full of diversity based on economic conditions, cast, religion, region, language, gender etc., there is a need of maintaining Data Warehouse in following areas [341].

9.2.1 Population Data Warehouse

The Data Warehouse should have various inputs representing diversity in India based on regional, religious, rural, urban, category wise, i.e. handicapped-visually impaired, hearing impaired, mentally retarded etc. Data Warehouse developed on the basis of such data collection at national level will be very

E Governance Data Center, Data Warehousing and Data Mining: Vision to Realities, 203–216.

effective through Data Mining for development of suitable strategies for population control.

9.2.2 Data Warehouse for Food and General Supplies

With regard to food supply to Indian community at large, with substantial population below poverty line, diversity in food uptake across the country, resources available for production of food grain and other eatables should be recorded and monitored. Such data through Data Warehouse could be used for estimating the production and consumption of food across the country to ensure quality food to every citizen in India. This Data Warehouse will also have appropriate consideration for present population and future projected population across the country [341].

9.2.3 Agriculture Data Warehouse

India has large number of farmers with marginal land. Considering the need for food, soil characteristics, input for fertilizers, pesticides etc., there are considerable amount of diversity in agriculture with regard to per acre turn over varying from rupees 5000/- to 150, 000/-. Hence, there is ample scope for agricultural planning in India. Despite a large group of marginal farmers as well as their inability to arrange for input, it is essential to develop Data Warehouse for all the attributes involved in agriculture. With appropriate Data Mining, a suitable strategy could be developed for agricultural production in which 74% of the Indian population is engaged. Consumption pattern of fertilizers, availability of various types of seeds, land use patterns, watershed management, pesticides etc., could be studied through suitable Data Marts for each of them [341].

9.2.4 Data Warehouse Based on Rural Development

As far as rural development is concerned, there are several government agencies of the State and Central Governments and certain non-governmental organizations working on various aspects of rural development including drinking water, microfinance and other programs. Therefore it is necessary to build appropriate Data Warehouse with provision for a suitable Data Mart for each activity of rural development. Considering a large number of individuals below poverty line, a suitable Data Warehouse for them is essential [341].

9.2.5 Health Data Warehouse

In health sectors, different types of data like community need assessment data, immunization data, and data from national programs on controlling blindness leprosy, malaria may also be used for Data Warehousing implementation and Data Mining Applications. Following predictions may be done in health sectors [341].

- Identification of noncompliant and fake activities, mainly in health insurance.
- Identification of unknown patterns related to clinical diagnostics or drug research.
- Identification of clusters from country population based on some common characteristics.
- Develop user profiles.

9.2.6 Planning and Development Data Warehouse

A Planning and Development Data Warehouse could be built for all sectors such as labour, energy, education, trade, industry, five year plan etc. The 6th All India Education Survey data has been transformed into a Data Warehouse (with about 3 GB of Data). Various types of analytical queries and reports could be answered by using this Data Warehouse [341].

9.2.7 Data Warehouse Based on Commerce and Trade

Considering the export and import in India for last 63 years especially after independence, the trade balance continues to be deficit while on the other hand China in last few years has altered the total position including trade balance in their favour. Obviously a suitable Data Warehouse for export and import is essential to develop so that their activities could be accelerated at par with other developing countries including Brazil and China. It is essential to develop short, medium and long range strategies for export so that our trade balance could become comfortable in next five years. For each sector of trade, there is a need of separate Data Mart [341].

Ministry of commerce is dealing with various foreign companies investing in India or doing business in India in some another ways. Export and Import is an important task which always involve large number of companies and high competiveness. Ministry of commerce is divided into seven exports processing zones and maintaining a Data Warehouse which maintains the data of each zone. It facilitates zonal wise roll up and drill down operation of daily data. It is also possible to view data in terms of week, months and on early basis by using aggregation. Various possible association regarding export/import parameters could also be explored by using this Data Warehouse. Following are the observation performed by the Data Warehouse:

- Monitoring System for Global pricing of the product.
- Comparative report on Export and Import status of different countries.
- Export/Import trade policies directions and trends.
- Global data bank related to Export/Import.
- Department level accounting, employee, operational details.

9.2.8 Finance Data Warehouse

The Department of Finance could utilize Data Mining applications as follows:

- Identification of noncompliant and false transactions/activities.
- Collection of user information.
- Associating different items and events for audit.
- Identification of essential data that affect target performance metrics.

9.2.9 Banking Data Warehouse

Banking sectors could utilize Data Mining applications for following purposes:

- Computing the probability of default (PD) for credit risk
- Bankruptcy prediction.
- To perform exploratory data analysis and pre-processing.
- Identification of different performance metrics and statistical based rules for borrowers.

9.2.10 Data Warehouse for World Bank

The World Bank maintains huge collection of data related to economic conditions, health, education and environment. The main task of the World Bank is to provide financial assistance to all its member countries. Here, it is important to maintain the profiles of various countries so that the correct and suitable country could be identified for financial support under certain parameters. For this purpose a Data Warehouse for World Bank is maintained which is identified as live database of World Bank. This Data Warehouse provides direct user interface for data access and includes various reports as required by government officials and economist. The Data Warehouse also analyzed the effectiveness of the financial aid which is provided to any country by the World Bank [341].

9.2.11 Data Warehouse for Department of Tourism

Department of Tourism is dealing with several thousands of visitor's everyday and significantly affects the reputation of any country [342]. A Central Data Warehouse is essential to store the record of every visitor so that not only the normal functioning but also various special securities, custom check could be performed at minimum time. Following are the expected outcome of a Data Warehouse to be maintained by the Department of Tourism:

- A secure arrival verification system for Foreign Tourist.
- Visitors' profiles/preference list with behavioral details.
- Department level earning expenditure details.
- Promotion of tourist destination/packages.
- Hosting of websites, employee database.

9.2.12 Data Warehouse for Tamilnadu States

In a state Data Warehouse it is essential to maintain data about the government operation, citizens and projects ongoing in all its district, village and block levels. This Data Warehouse serves quick government decision making for administrative, investment and social welfare purposes. This Data Warehouse includes various Data Marts on health, agriculture, weather, basic amenities for various decision making processes. It is a web based interface so that a user friendly environment is available for viewing reports in above mentioned areas [341].

9.3 Data Mining for Indentifying Countries for Financial Support

For any financial support it is crucial to identify few countries on the basis of certain indicators. Generally the factor of good governance, status of human development and effectiveness of financial support are three important parameters and deciding factors for financial support of a country. Variables that have been taken to identify the correlation among good governance, status of human development and effectiveness of financial support are mentioned in following table [343]:

Table 9.1 Lists of Variables for Calculating the Correlation Among Good Governance, Human Development and Effectiveness of Financial Support

Variables	Description
Voice and Accountability (VA08)	Freedom of citizens to express and to participate in choosing government.
Political stability and absence of violence (PV08)	Probability that a government will be overthrown.
Government effectiveness (GE08)	Quality of government services and how independent they are from political pressure.
Regulatory quality (RQ08)	Capacity to design and run policies and regulations that foster private enterprise.
Rule of law (RL08)	Ability to enforce laws through the police and courts.
Control of corruption (CC08)	Extent of using public power to advance private interests.
Official development assistance (ODA)	A rank for Financial support
Average & Governance Indicator (GIAve)	Average Governance Indicator calculated on the basis of above parameters

The details of countries along with the variables which are used for the analysis are indicated in Figure 9.1:

A decision tree has been prepared to identify the influence of different variables for deciding different class levels .This decision tree as shown in Figure 9.2, clearly classify the countries in three groups - high, medium and low chances of getting the grant. The country having low political stability and absence of violence with high rule of law is belonging to a class having highest possibility of getting financial support. The leaf TOP (31.0/12.0) indicates that this model correctly classified only 31 countries in the high groups while making error in 12 of them. The country for which PV 08 is greater than 0.06 and RV 08 is more than -0.43 is having the low chance of getting the grant [343].

9.4 Data Mining in Security Perspective

The project Total Information Awareness (TIA) was instigated by the US Government subsequent to the terrorist molest of September 11, 2001. The intention of TIA was to search large data and determine associations and patterns connected to terrorist activities. The project conducted discovery of associations among transactions such as work permits, credit card, air travel

	VA08	PV08	**GE08**	**RQ08**	RL08	CC08	**GI Ave**	ODA million $
Iraq	-1.26	-2.69	**-1.41**	**-1.09**	-1.87	-1.48	**-1.63**	9.114.71
Afghanistan	-1.26	-2.64	**-1.31**	**-1.58**	-2.01	-1.64	**-1.74**	3.951.08
Tanzania	-0.09	0.01	**-0.45**	**-0.39**	-0.28	-0.51	**-0.29**	2.810.84
VietNam	-1.62	0.32	**-0.31**	**-0.53**	-0.43	-0.76	**-0.56**	2.496.73
Ethiopia	-1.3	-1.79	**-0.43**	**-0.86**	-0.6	-0.66	**-0.94**	2.422.48
Pakistan	-1.01	-2.61	**-0.73**	**-0.47**	-0.92	-0.77	**-1.09**	2.212.42
Sudan	-1.77	-2.44	**-1.41**	**-1.36**	-1.5	-1.49	**-1.66**	2.104.19
Nigeria	-0.6	-2.01	**-0.98**	**-0.62**	-1.12	-0.92	**-1.04**	2.042.33
Cameroon	-1.02	-0.53	**-0.8**	**-0.66**	-0.99	-0.9	**-0.82**	1.932.60
West Bank & Gaza	-0.94	-1.76	**-1.36**	**-1.12**	-0.81	-1.13	**-1.19**	1.868.20

Figure 9.1 List of Countries Along with Different Variables

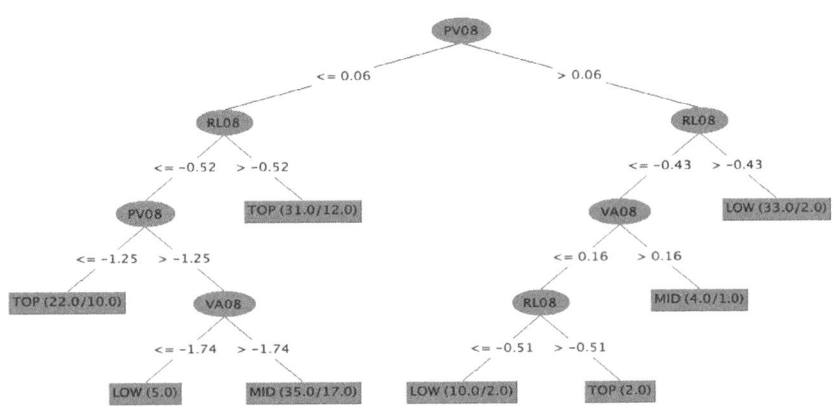

Figure 9.2 Decision Tree Classifications for Distribution of Financial aid

tickets, passports, visas, driver's license and trials such as arrest or doubtful activities [344].

CAPPS is Computer Assisted Passenger Pre-Screening System developed by US Government in the year of 2004. CAPPS is a prescreening arrangement. It is implemented to check all airline passengers against a commercially available database. After checking, it provides a risk color or status to each passenger. CAPPS collect information provided by the passenger for example name, contact details etc. These records are then forwarded to commercial data providers for passenger assessment and correlation with other events. CAPPS was conducted in two phases CAPPS I (2001) and CAPPS II-b(2003) and later associated with an another plan to discover persons who are prohibited inhabitants of the country for example persons with expired visas, prohibited aliens, etc [344]. The CAPPS program was limited to aviation security only and had been condemned in terms of data integration storage, access and reliability related problems.

In August 2004, another new project SECURE FLIGHT was started in the place of CAPPS II due to high cost and troubles related to project management [344]. It was again concerned about aviation security and contained a Blacklist data of suspicious passenger. This dataset was matched with passenger's record and an association analysis is conducted to generate "No fly list" and "Selectee list" to indicate a passenger clearance for travel. The Secure Flight was further suspended in 2006 because of system development and management related problems.

In May 2004, a report indicates that there were around 200 Data Mining programs started by government which does include any secret Data Mining activities. Figure 9.3 shows most listed application areas.

The MATRIX collects information regarding citizens from different government and private databases. The MATRIX pilot project was based on an automatic analysis system works as a technological, exploratory and allows searching of existing records in data repository [344]. With the help of this application a vast search operation could be performed. The search operation uses various combined and disparate datasets related with citizen and then assembles the results. It calculates High Terrorist Factor (HTF) based on which the identification of terrorist could be done [345].

A project Able Danger was initiated in 2004 by the U.S. Army. It was aimed to combat global intimidation. It was based on link analysis to recognize interconnection between persons who otherwise emerge with no noticeable link with one another. Figure 9.4 shows the link analysis based Data Mining. It was performed with 2.5 terabytes of data collected worldwide. Later it was found in notice that "Mohammed Atta" one of the attackers of 9/11 was identified as a result of Data Mining operations.

Data Mining with appropriate Data Warehouse could be used to slowly decrease and finally eradicate the undesirable activities in the society. This approach is essential for slowly reduction and eventually elimination of terrorism by analyzing the behavior of the terrorist, their social and family background,

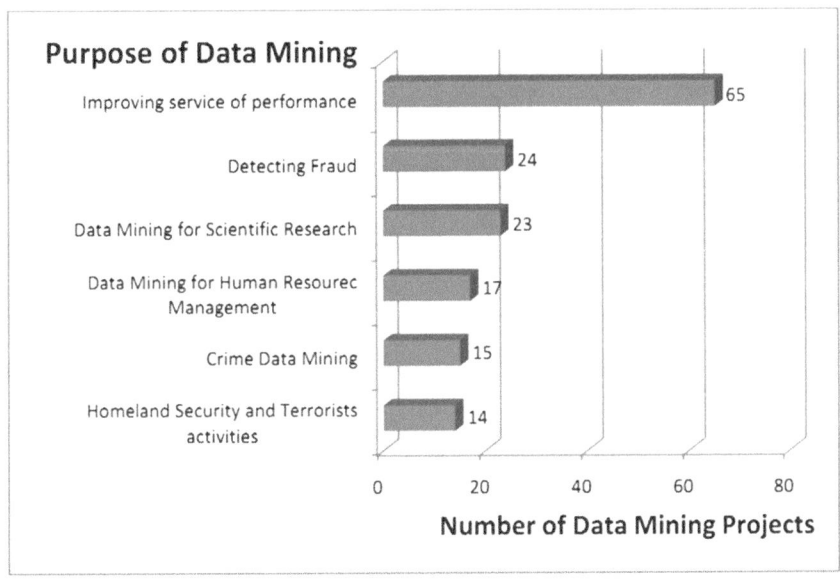

Figure 9.3 Data Mining Application Area

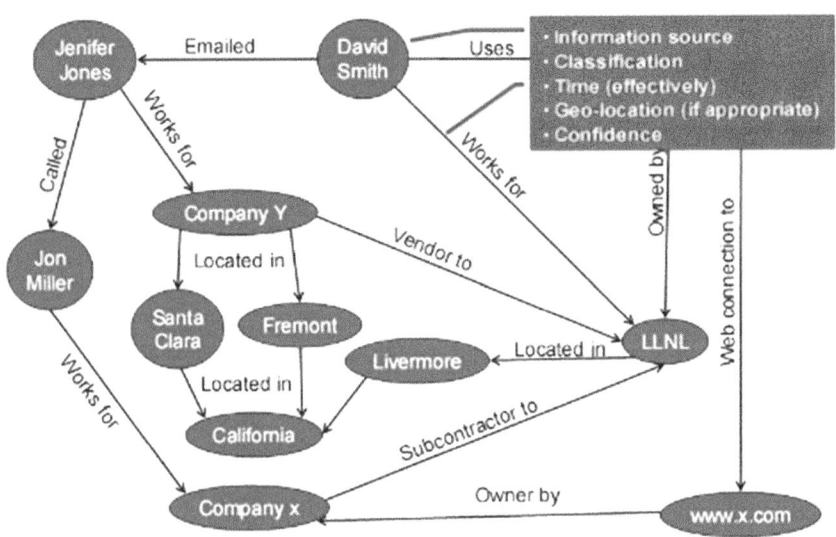

Figure 9.4 Data Mining Using Link Analysis

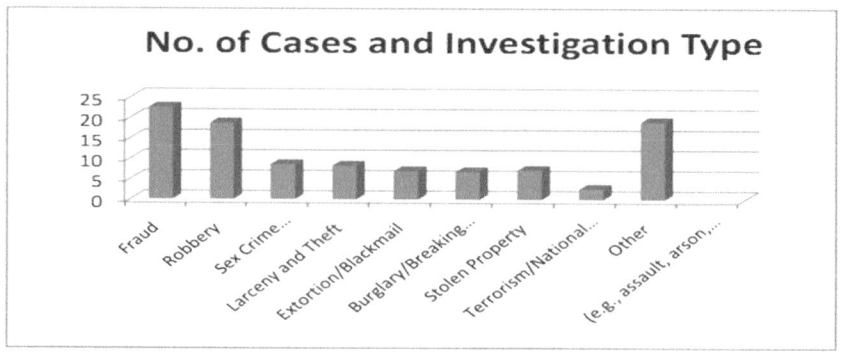

Figure 9.5 Number of Data Mining Cases and its Investigation Types

Table 9.2 List of System Used in Security and Counter Terrorism

System/Component	Description
Analytical Framework for Intelligence (AFI)	Customs and border protection (CBP) developed this system to collect data from various border sources and access this data from single interface and enable a system analyst for searching and discovering pattern.
Automated Targeting System (ATS)/ATS-Passenger (ATS-P)	CBP works on pattern based approach. It analyzes data and information for targeting purpose. This data is useful in preventing terrorist entry and weapon entry in United States.
Citizen and Immigration Data Repository (CIDR)	U.S. Citizenship and Immigration Services (USCIS) use this system for preventing fraud in immigration information. It responds to request comes from DHS office
Data Analysis and Research for Trade Transparency System (DARTTS)	It uses to investigate import and export crime. It also include crime such as trafficking of illegal goods, money laundering etc.
TECS/TECS Modernization (TECS-Mod)	It deals with data related to traveler's entry/exit information for preventing unauthorized entry from border.

education, potentiality for employment etc., over a period of time by means of appropriate Data Mining for strategic management of such undesirable activities. Figure 9.5 shows number of Data Mining cases and its investigation types.

Following are some other data mining application in security and counter terrorism:

9.5 Data Mining Techniques for Judicial System

Data mining algorithms have been used to construct computer based knowledge and widely applicable in all domains. In judicial system Data Mining application is gaining popularity nowadays because of ever increasing size of legal databases. For well-organized implementation of Data Warehousing and Data Mining in judicial system, primarily, it is important to develop an information system. This can minimizes the time and effort in the present judicial work which is complicated in nature. The current judicial system is lacking with no consistent and standard formats, scattered online case resources and random schedule.

By using an automated system accurate and appropriate cases can be found quickly by the judges. It helps to save judicial resources and in turn improve judicial efficiency. Data mining concept will be used to access the information system efficiently and use the knowledge of the results announced previously on similar cases in future. In judicial system any present case requires the rules and regulations to be referred from previous cases so a case repository system to be developed and continuously updated [346]. This could greatly help to enhance learning efficiency, update knowledge of law and reduce searching cost of reference cases and legal knowledge.

A Judicial system with Data Mining consist various modules according to the functioning of a legal system. Here XML standards are used for document structuring and it is designed to work like a thought process of a judge. The system also assists the judges in faster information access, well-organized retrieval system for legal laws and reasoning [346] [347]. Major task of Data Mining algorithm in judicial system are as follows:-

- A classification step is required to classify the criminal element and non criminal element by using a crimer parser.
- A mapper is required to map the criminal acts to their respective punishments as per law.
- It finally identifies the judgements which contain punishments.

According to the above Figure 9.6, the approach consists a training phase for training of certain rules indicated by dashed arrow and a testing phase for further validation indicated by normal arrow style.

9.6 Public Opinions Mining in Governmental Decisions

E-Government exploits the advancement of Information Technologies to increase the quality of services and information offered to people, to establish

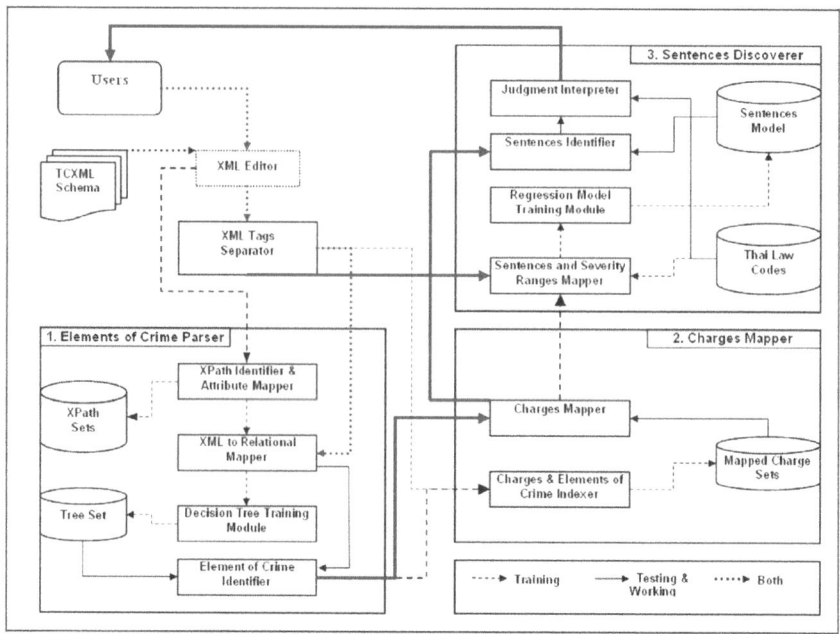

Figure 9.6 XML Standard Diagram in Judiciary System

accountable and transparent governance for public. One of the most important concerns for successful E Governance is to encourage citizen participation in governmental decision-making process. Internet is playing significant role in citizen participation because it provides connectivity to people to communicate with government [348] [349]. Through internet Government organization can also publish all their decisions and documents online.

To realize the potential of internet in citizen participation and E Governance, it is needed to integrate Text and Data Mining techniques for modeling the public opinions and assessing the government decisions. Following are the steps of opinion mining in government decision making [348]:

- Firstly identifying the public opinions (communicated online) about government decisions by using Text and Data Mining.
- Secondly analyzing the effectiveness of the mined opinions therefore it will be considered in subsequent government decisions.
- Lastly explore the correlation between mined opinions and the formulation of new decisions for future reference.

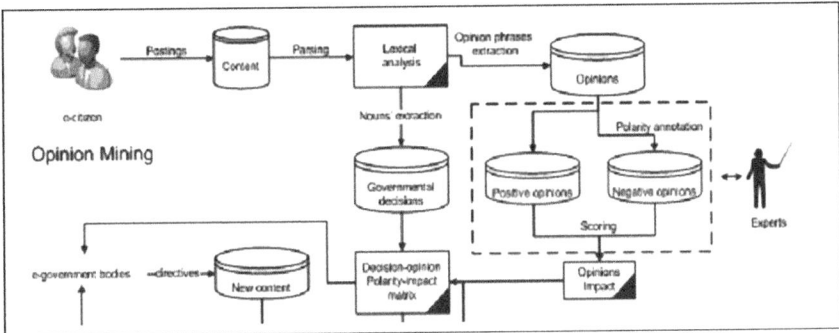

Figure 9.7 Opinion Mining Framework for Government

Figure 9.7 represents the opinion mining framework for government. Initially all users online comments regarding any policy decision are considered. Then a parser and lexical analyzer are used to detect and retrieves phrases containing user opinions from their online posts. After extracting all the opinion a polarity annotation techniques as classifier is used to differentiate between positive opinion and negative opinion [348] [349] [350]. Then, for every government decision for which there are some public opinions articulated, two data clusters are required to be generated:

- In positive cluster, all adjectives which are used to support the government decision are placed.
- In negative cluster, all adjectives which are used to oppose the government decision are placed.

This method provides a framework for government as well as public interaction based on opinion mining which can efficiently control, recognize and authenticate public opinions automatically.

9.7 Conclusion

At present Data Mining is not in use for development of strategic planning and control in social welfare activities in India at local level, state level as well as central level. Obviously Data Warehouse, which is well developed and maintained for all such social welfare and developmental activities, is not available. This has been one of the serious constraints for thorough analysis through the techniques used but, once Data Mining is recognized as a meaningful tool for strategic planning and control, appropriate Data Warehouse for each such activities related to social welfare and development will be maintained for Data Mining Analysis. These different level of Data Warehouses/Data Marts should be integrated together to exchange information. This secure data can be further utilized by all the citizens of India and various Planning and Development, Regulatory and Social Welfare Departments of the country. The approachability of the Data Warehouse for the common man is important and can be insured using independent computerized embedded systems like KIOSKs. The proposed Centralized Citizen Data warehouse may be further imposed on various Data Mining techniques like Classification, Regression and Clustering. Theses mining algorithms may predict various important rules based on observation of previous historical data. There may be following predictions:

Imbalance in Gender Ratio: Gender ratio is the ratio of male to female newborns. A Centralized Citizen Data Warehouse and efficient Data Mining may outline immediate changes in the gender ratio. These trends can be linked with other social circumstances and to predict a final rule for meeting the gap in Gender Ratio.

Rate of Birth /Death and Population: Data Mining may also help us in identifying the root causes of population growth and relationship between Birth and Death Rates. For example, a Data Mining analysis of Meghalaya identified that although both the birth and the death rates have come down over the years, the gap linking these two has not shrunk as per expectation. This imbalance could be considered as one prime ground for population explosion in the State. Another observation stated that population also increases because of excess of immigration over emigration.

Career and Education: Data Mining and Warehousing can utilize large education databases and predict success rules for students which may help them choose suitable courses.

Reduction of wastage of government resources: Data mining algorithm can classify datasets into similar groups based on experiences. Suppose a charity wants to communicate with different donors in case of Emergency, using Data Mining can help examine the characteristics of past donors and hence reduce mailing list and wastage of resources.

10

Data Warehousing and Data Mining Applications in Different Sectors

10.1 Introduction

Data Mining and Data Warehousing is an interdisciplinary process having broad range of streams such as Health, Finance, Business and Telecommunication etc. All these organization presently believe that organizational data must be collected, processed and stored in well defined manner in order to produce complete, reliable and good quality data which is the prime requirement of any Data Mining initiative. Data Mining is essential and powerful technology which helps various industries to improve their business practices and avail competitive advantages. A Data Mining application is typically associated with a Data Center, Data Warehouse and various techniques of knowledge discovery in databases for instance clustering, classification, regression or association rule mining. This chapter illustrates Data Mining applications in Health, Finance, Banking, Insurance, Business, Education, Agriculture and Telecommunication etc.

10.2 Data Mining Applications in Healthcare

In healthcare the growth rate of electronic records has been progressing fast and creating the need of Data Mining and intelligent data analysis techniques. Data Mining also becoming popular because it is advantageous to doctors, patients, health centre and researchers [351]. It includes the assessment of treatment success rate, management policies, patient satisfaction and also the detection of fraud and abuse. Figure 10.1 illustrates various stages of Health Data Mining. In healthcare the Data Mining can predict the patient behavior on the basis of event based guidelines and improve the treatment process. It collects data from various sources related to medical domain and utilizes healthcare repositories or Data Warehouses for information extraction which

E Governance Data Center, Data Warehousing and Data Mining: Vision to Realities, 217–236.

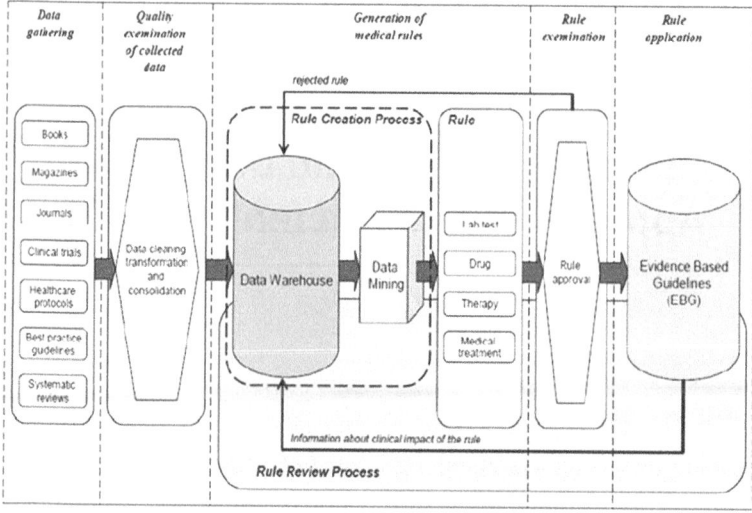

Figure 10.1 Phases of Health Data Mining

further becomes knowledge and used to improve the efficiency of patient care. On the basis of certain rules and suitable prediction algorithms, it significantly reduces patients' risk and diagnosis costs.

Lots of unsupervised algorithms are also there to find the natural clusters of patients. Through Data Mining, the physician can obtain clinical rules on the basis of past profile of various patients. It can also find the hidden patterns in large Data Warehouses. Figure 10.2 explains the functioning of a Data Warehouse for Data Mining.

Following are the promising advantages of Data Mining in Healthcare:

10.2.1 Assessment of Treatment Success Rate

Data Mining is widely used as an assessment tool for determining the treatment success rate. It considers causes of disease as attributes and discovers some interesting patterns. For example, a comparative analysis may be done to evaluate different treatment plans for any particular disease and the most effective plan may be explored. In this manner various clinical profiles could be developed for further standard guidelines regarding diseases. Some other Data Mining applications in healthcare includes correlation between disease, drugs and their side effects and determining most common signs regarding any disease [351]. Effectiveness of various drug compounds for different

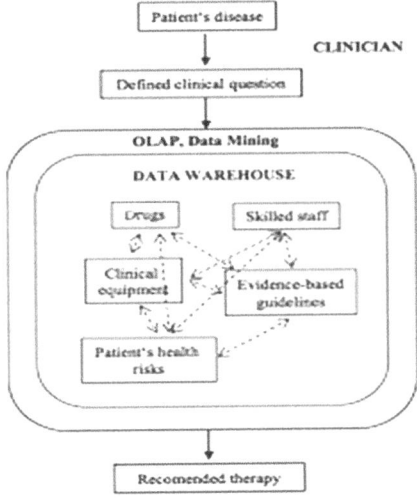

Figure 10.2 Functioning of a Health Data Warehouse

categories of patients may also be identified through Data Mining. Figure 10.3 explains a commonly used Association Rule based Data Mining approach to predict possibility of infections in patients. Similarly, Figure 10.4 classifies the group of patients as low risk patient and high risk patient.

10.2.2 Healthcare Resource Management

Managing hospital resources is one of the most important tasks in healthcare that facilitate maximum utilization of hospital facilities for example medical claims, hospital admissions etc. In order to utilize hospital facilities properly, there is a need to detect chronic diseases in accurate manner and prioritize the treatment of patients on the basis of criticality of disease [351]. Demographic conditions and fitness report of a person is also plays important role in optimum utilization of available hospital resources.

10.2.3 Fraud and Abuse

Data mining establish a model to identify any fraud and abuse in medical claims. This model can easily detect the previously unknown or irregular patterns in the medical claims, highlight improper prescriptions and fake insurance claims made by patients, physicians, hospitals, laboratories etc [351]. In this manner, it is possible to enhance annual savings up to greater extent by

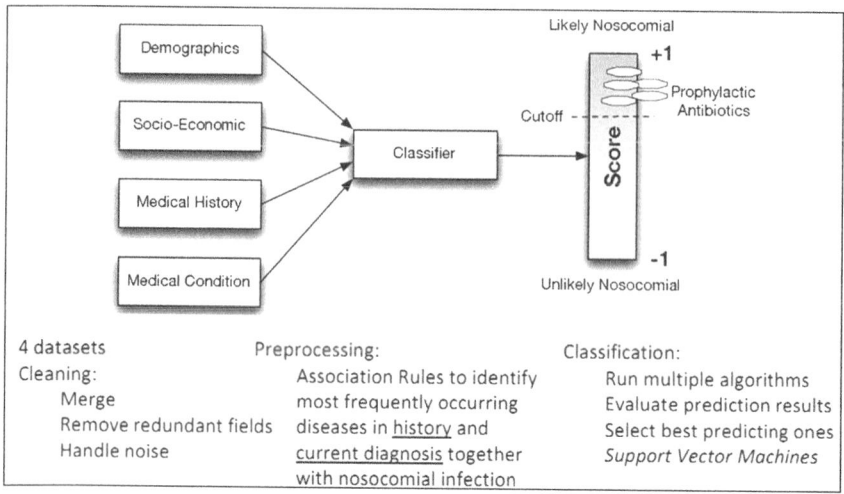

Figure 10.3 Association Rule for Infection Detection

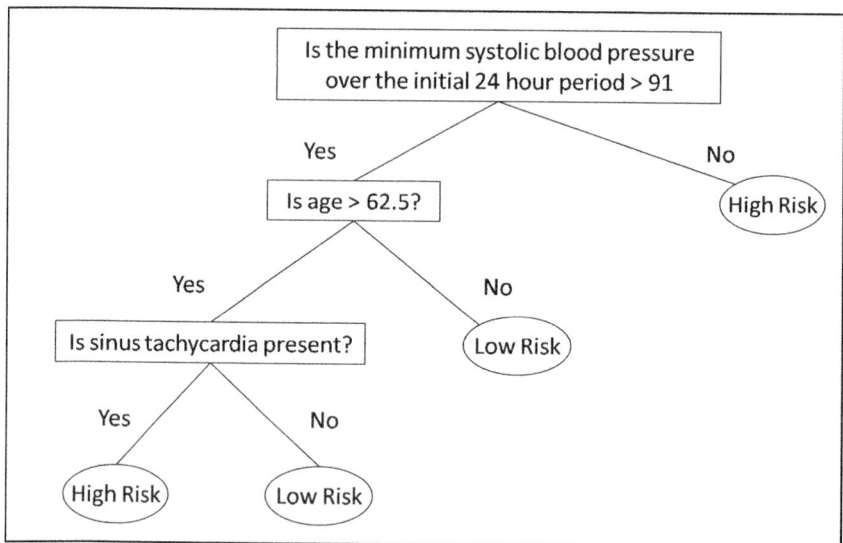

Figure 10.4 Decision Tree Classifier for Patient Data

discovering the unusual patterns and fraud from the huge data obtained from millions of operations, referrals and treatment courses.

10.2.4 Recognize High-risk Patients

Various companies have been using Predictive Data Mining approach to recognize the patients of high risk. This system provides support to the hospitals to manage all types of diseases, improve the health quality of the patients and reduce the treatment cost of the patients [351]. Using Data mining techniques, a predictive model is developed that helps the healthcare providers to identify high risk patients and provide effective and appropriate services to the patient.

10.2.5 Grading Hospitals

Based on the information which is obtained from different healthcare providers, organizations may generate the rank of hospitals. Various Data Mining approach such as classification, clustering and association are used to analyze the reports of the hospitals [351]. On the basis of these information any hospital that does not consider the risk factors seriously, may be allotted lower rank even if its success rate is equal to other hospitals because it has high death rate. Thus, Data Mining techniques facilitates comparison among different hospitals on the basis of their capability of handling high risk patients.

10.2.6 Hospital Infection Control

Data Mining techniques for surveillance system are used to discover unknown or irregular patterns in infection control data of any disease. The system may use association rules to produce interesting pattern from the data obtained from laboratory information management system. These patterns may further reviewed by an expert to control the infection in hospitals [351] [352].

10.2.7 Enhanced Patient Care

Data Mining helps the healthcare organizations to identify the present and future requirements of patients and their preferences to enhance their satisfaction levels [351]. For example, a patient healthcare usage index using Data Mining techniques may give suggestion about a person's tendency towards the usage of particular healthcare services. In this manner a recommendation system has been used to provide the appropriate services to the right patients at right time. It helps to determine the requirement and expectation of a particular patient in promising way. It is also helpful to encourage people regarding various disease prevention techniques. Using Data Mining techniques, different Pharmaceutical companies can determine the response of patients regarding drug therapies. These companies can easily track which drugs is prescribed by which doctor and for what purpose.

10.3 Data Mining in Banking

Automation of banking system and other financial institutions are responsible for huge data generation. Data Warehouse of such organizations contains four categories of data and information i.e. Financial, General, Individual and Corporate. Data Mining consist a set of machine learning and statistical models for quick decision making and behavior prediction through classification, clustering and regression. It also helps banks to improve their understanding with their customer so appropriate marketing campaigns could be planned [353]. Following are the significant advantages of using Data Mining in Banking System:

10.3.1 Marketing and Customer Care

Data mining is used to enhance customer services in responsive manner especially with the help of Data Repositories. This could accelerate overall performance with low cost. With the help of Data Mining history of previous transactions could be referred to decide future strategies and policy decisions [353] [354].

10.3.2 Customer Profiling and Relationship Management

A recommender system based on Data Mining is an emergent application, used to support customers as well as policy makers for their quick decision making. In banking system customer profiling is required to identify reliable customers so that their possible behavior could be predicted [353]. For example, any customer may be considered as "risky" or "safe" on the basis of his/her historical patterns. Similarly the suitable amount of loan for any customer could be easily predicted by using regression. Customer retention rate can also be well estimated and improved by using historical knowledge about the customer.

10.3.3 Trading

In trading, predicting temporary movements of currency exchange rates, interest rates, equity values are essentially required. These fast changing factors are

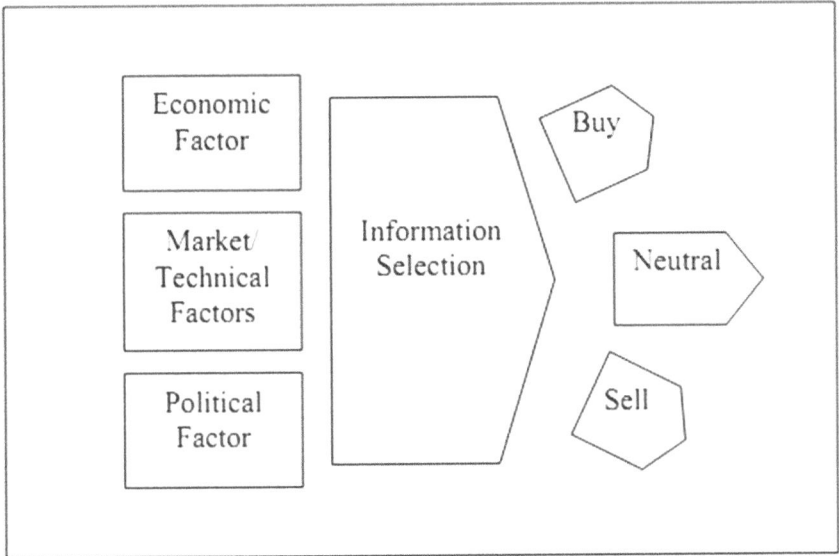

Figure 10.5 Influencing Factors for Trading

based on market, economic as well as political conditions and affect various buying/selling activities. Figure 10.5 illustrates that if appropriate input has been given in terms of economic, market and political factors, a Data Mining tool can easily predict suitable time for trading which includes profitable buying and selling opportunities [353].

10.3.4 Risk Management

Risk Management is very crucial in banking systems and other organization such as Insurance etc [354]. At present it is imperative to consider two prime risks i.e. market and credit risk, mainly depending upon customer, instrument and portfolio risk as indicated in Figure 10.6.

Market Risk consist various financial market parameters such as Stock indicators, Bank interest rates and value of currencies which are important and changing frequently. Data Mining could help a market expert to determine their changing pattern so that the same may be utilized in various trading processes. Credit Risk assessment is essential in commercial lending process. It is used to identify the right customer for safe lending so that the loan should be realized as per given time [355] [356].

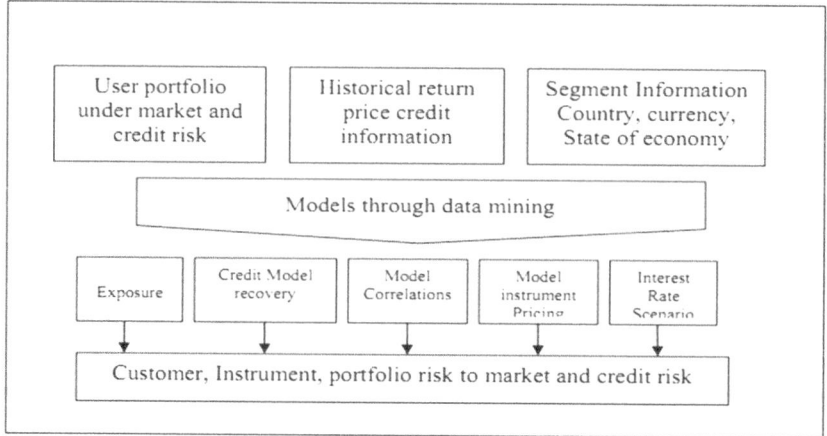

Figure 10.6 Various Types of Risk

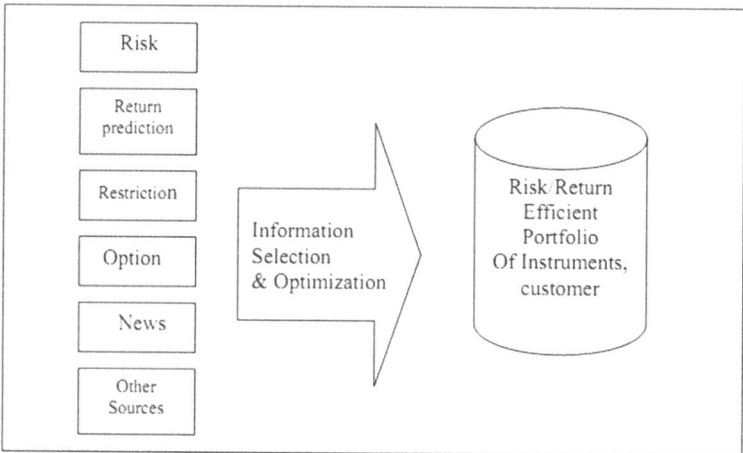

Figure 10.7 Portfolio Management for Risk Management

10.3.5 Portfolio Management

Portfolio management is intended to estimate the risk levels involved with any customer or product. It mainly involves profit/loss analysis and return/risk analysis as per changing market condition. A Data Mining tool can help investor to identify most profitable trading deals with minimum risk [353]. The Figure 10.7 shows that a portfolio management is depending upon various sources and essentially used in risk profiling.

10.4 Data Mining in Finance

Major banking and financial organizations offer broad range of services i.e. savings, loans, credit limits, long term and short term investments. Data mining techniques are used for systematic analysis of financial data that is obtained from financial and banking organization so that performance, accuracy and satisfaction level could be achieved in greater extent. The different application of data mining in finance is given below [357]:

10.4.1 Building a Data Warehouse for Analyzing Data Form Multidimensional Point of View

Financial data are fluctuating in nature having several dependencies upon fast and frequent changing economical, legal, technical and social scenarios. The banking and financial data warehouse is needed to facilitate multi-dimensional view of finance data so that all existing correlation and dependencies must be clearly identified and treated with fair consideration [357]. In multi-dimensional view quick observation and decision making is possible to obtain for example a single query can answer a customer his/her income and expenditure pattern that varies by month, sector and region. Multivariate Data Warehouse techniques like data cubes along with its roll-up and drill-down operation can able to answer dependencies, discovery of hidden patterns and outliers for all complex financial analysis.

10.4.2 Analysis of Customer Credit Policy and Loan Payment Prediction

For successful banking know your customer (KYC) is essentially required so that the credit risk as well as loan risk could be avoided without any loss. There are following factors which affect the customer credit rating and loan payment:

- The ratio of payment-to-income and loan-to-value.
- Level of customer education
- Period of loan
- Income of the customer.
- Balance Details.
- Previous record of customer credit.
- Residence Area

Further studies in relevant field indicate that among all attributes of financial data some are highly influencing and few are least influencing. Data Mining helps to perform attribute selection for effective results. For example in above mentioned factors the ratio of payment-to-income is having high influence for making loan payment policies while other factor such as educational background and balance details is not so important.

10.4.3 Data Mining for Targeted Marketing

Various Data Mining techniques such as classification and clustering are used for discovering the customer groups for focused marketing [357]. Different clustering techniques are implemented for making group of those customers that have similar behavior or practices regarding banking. Effective data mining approaches can help to identify the pattern of the customer groups. It also associates a new customer with that certain customer domain to facilitate the marketing strategy and business modulation.

10.4.4 Data Mining in Financial Crimes

Significant information from various databases such as bank transaction or crime databases are collected for discovering the financial crime such as money laundering etc. Various data mining techniques are used for the identification of unusual or unknown patterns for example large quantity of cash laundering in a particular interval by a certain groups of people etc. Other data mining tools such as data visualization, classification, clustering, outlier analysis, linkage and sequential pattern analysis are used to detect the important association among different activities which in turn helpful for investigating doubtful cases [357].

10.5 Data Mining in Business

Data Mining is playing leading role in business applications for enhancing business intelligence, performance, reliability and sustainability. In sales analysis Data Mining explores associations between related products such as milk and bread, strawberry pop tarts and wine which are interesting and helpful to determine selling policies. Similarly in Customer Profiling, Data Mining can advise purchase habits of different products which are helpful in target marketing. Appropriate recommendation could also be obtained to attract new customers on the basis of hidden patterns existing in historical database. Following are the established Data Mining application in business:

10.5.1 Customer Segmentation

For effective marketing of retail organization, it is necessary to partition the customers into different groups on the basis of their characteristics. These segmentations can provide an insight view of customer behavioral pattern on the basis of different demographics [358], seasons and social circles in following manner:

- Reaction of the customers for discounts.
- Reaction of the customers for newly launched product.
- Response of the customers for new promotions.
- Purchasing behavior of customers for particular products.

10.5.2 Campaign/Promotion Effectiveness Analysis

For successful launching a business campaign it is important to plan its target segment, media coverage, cost overhead and expected outcome. A Data Mining insight in advertising greatly facilitates good market coverage with the help of following parameters [358]:

- Successful media channels for previous campaigns.
- Responses of different groups of customers to the campaign.
- The comparative costs and benefits of the campaign.
- The geographic location at which that particular campaign is launched.

10.5.3 Customer Prioritization

It is not possible to provide equal profit to all customers at the same time. So, it is extremely necessary to prioritize prime customers for special packages to establish long-term relationship with them. Appropriate customized models may be constructed for customer prioritization with the help of Data Mining techniques [358].

10.5.4 Customer Loyalty Analysis

Various studies have shown that maintaining existing customer is more economical and challenging process as compare to acquiring a new one. To establish an efficient customer retention strategy, it is important to determine the reason for customer abrasion. Business intelligence formulation through Data Mining assists to understand the customer retention rate [358].

10.5.5 Cross Selling

The Business Management groups examine the huge amount of customer information available in a retail store to sell other products together. This methodology works on the basis of previous purchase history of any customer to analyze their shopping interests using Business Intelligence tools [358].

10.5.6 Product Pricing

Deciding pricing tags is most significant marketing agenda in retail. Retailers can develop a cost model using Data Mining. On the basis of the cost of different products a price-to-sales association could be determined for establishing a profitable combination of price/sales for a particular product [358].

10.5.7 Target Marketing

Retailers and Business Intelligence providers can optimize the global marketing resources and promote new products by commercial campaigns. Target marketing practices are base on the buying habits of customers or some certain customer groups. Similarly Data Mining tools segments specific customer groups which respond to particular range of campaigns [358].

10.6 Data Mining in Insurance

Insurance industry deals with all types of customers, needs to maintain their reliability, sensitivity and accountability [359]. Data Mining could play a vital role in this sector because it is dealing with numerous uncertainties depending on economic, political, social changes of the society. Following are the key advantages identified in insurance industry using Data Mining:

10.6.1 Customer Profitability

Identifying the most profitable consumer or the group of customers from huge dataset of retail management involve several steps [359]. To identify most profitable deal with a consumer, the business insurers must quantify the following factors for high accountability:

- The expenditure on customer services over a period of time.
- The profits obtained from the customer.

For example a customer may not be profitable because he/she may be using a product that does not match with their risk profile. Customer profitability analysis can efficiently support the marketing strategy development process for a fresh product as well as existing products.

10.6.2 Customer Segmentation

Segmentation process is used for dividing the customers into different groups based on their characteristics. Data mining techniques such as clustering is used to segment the customer into different cluster in such a way that the customer within segment has similar behavior while the customer belongs to different segment exhibit different characteristics. Grouping the customer is done on the basis of various factors such as demographic, cultural, social and psychographic [359].

10.6.3 Attrition Analysis

In retail Industry it is clearly established that obtaining new customers is more expensive than maintaining the existing customer. Insurance related products demands long-term association between consumer and insurer [358]. Any pre-mature closure involves losses for both parties thus it is important to have fair customer relationship management policies for retention rate. A Data Mining analysis may help to avoid cases of pre-mature closure as well as also highlight product promotion policies by using correlation, regression and association rule mining.

10.6.4 Affinity Analysis

Certain product items contain a common relevance with each other and expected to be bought together [359]. For example, if a customer is purchasing an insurance policy in early 30, there may be a chance to seek certain allowances at the same time. These profitable resemblances in the insurance industry are very difficult to discover. So for this purpose Data Mining tools are used to identify suitable combinations of insurance products.

10.6.5 Campaign Analysis

This process is used to examine the impact of product promotion operations on the society. Data Mining tools are used to analyze the outcome of the particular campaign. The information of campaign analysis is recorded and stored in a Data Warehouse where an impact analysis is performed for the campaign [359].

10.6.6 Cross Selling

It is the main source of income in insurance organization where customer database is utilized in versatile manner especially for Data Mining operations to identify profitable results [359].

10.6.7 Risk Modeling

A predictive model using Data Mining techniques are constructed for determining the risk associated with individual profiles using various risk measurement issues such as claim frequency, amount of claim and loss ratio. For example, a rich person with frequent drinking habit, drives the sports car regularly may be considered as high risk customer [359]. A variety of risk measure parameters are calculated for particular customer group to estimating the right premium return.

10.6.8 Reinsurance

A reinsurance organization can calculate the insurer's risks factor against the return of the premium. Data mining techniques are used for developing the predictive model which helps the reinsurance organizations to determine the profitable reinsurance level [359]. Data Mining can also help actuaries to decide the suitable amount of the reinsurance in order to provide maximum benefits to the insurance organization.

10.6.9 Profitability Analysis

Data Mining techniques are used to analyze the profitability of particular product on the basis of various factors such as geographic regions, customer segment, organization, product line etc. [359]. A predictive model is constructed using data mining for determining the market value of newly launched product and identifying the targeted customer for this particular product.

10.7 Data Mining in Agriculture

Data mining in agriculture is an ultramodern research area consisting decision making techniques for agricultural domain. Recent technologies in data mining are capable to provide huge information on agricultural-related activities, which can be further to used in policy decisions. Following are few advantages of using Data Mining in agriculture:

10.7.1 Prediction of Wine Fermentations

Prediction of fermentation of wine would help the process to provide a regular and smooth fermentation level [360]. It also characterizes the farmers according to their family income, household size or on the basis of agricultural assets. It also helps to determine the pattern between crop yield and consumption of farmers according to land size.

10.7.2 Data Mining in Animal Husbandry

The early detection of the diseases could be mined using acoustic dataset. A computational system can be developed based on data mining approach which would be capable to monitor the sound of pig to discriminate the abnormal health pattern [361]. It also determines the cluster of the families based on their income level so that subsided supported may be given by the government.

10.7.3 Data Mining in Pesticide Usage

Data Mining in agriculture especially for pesticides usage could help to determine following issues [362]:

- Correct time for pesticide Spray
- Selection of right/wrong pesticides
- Temporal relationship between pesticide usages in days

10.7.4 Data Mining for Soil Profiling Prediction

Data Mining technique mainly classification is used to identify soil profiling. The model has been developed to establish patterns and correlations between soil properties and crop yielding, which would be beneficial to the farmers [363].

10.7.5 Data Mining for Predicting the Effect of Climate Change

Various data mining approaches are used to classify weather patterns. Parameters such as maximum temperature, minimum temperature, rainfall, evaporation and wind speed are considered to detect the relationships between crop production and climate change [364].

10.8 Data Mining in Education

Education Data Mining is an uprising research area highlighting methodologies for exploring data generated by educational organizations to understand

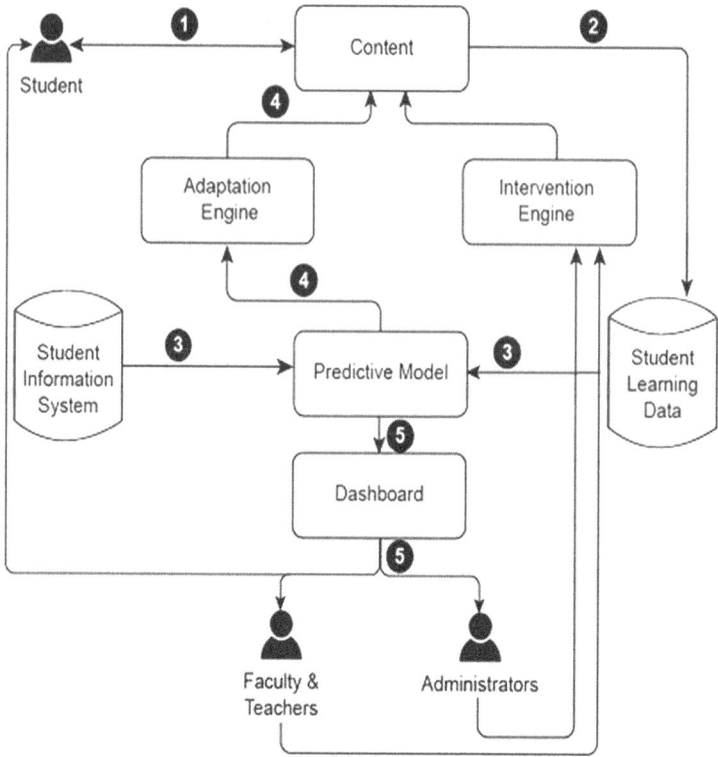

Figure 10.8 Education Data Mining

the students' performance and behavior pattern. The basic usages of Education Data Mining are prediction of student performance, relevant recommendations, course suggestions and enrollment data mining. Figure 10.8 explains a model for Education Data Mining which consist student information system, student learning data to support a predictive model and generates beneficial outcomes for students, faculty, teachers as well as administrators.

Following are the remarkable outcomes of Education Data Mining:

10.8.1 Data Mining for Improving Student Models

The application of Data Mining can discover the hidden patterns about a student's characteristics based on knowledge, meta-cognition, motivation, as well as psychological attitudes [365]. To understand individual profiles of the students using Data Mining is feasible in following manner:

- A model could increase the ability to predict student knowledge and their future direction.
- A model is used for guessing and predictions of future performance of the students.
- A Data Mining technique could assist to study the factors which lead the students to make specific alternative in their learning phase.

10.8.2 Data Mining for Learning the Pedagogical Support

The success factor of the students is depending upon his/her pedagogical background possess right from the beginning [366]. Data Mining may predict the impact of pedagogical background in the students learning efficiency. This Data Mining dependent factor indicates how each type of pedagogical supports is improving the learning system effectively. It could also suggest alternative methods of teaching for any specific group of students so that overall performance could be enhanced.

10.8.3 Data Mining for Human Judgment in Education

Most of the method in Education Data Mining is based on data visualization. For course selection, student opinion data could be preprocessed, displayed and compared with some already existing patterns to develop a predictive model [367].

10.8.4 Data Mining for Monitoring of Student Performance

The prediction of learning performance of students has been analyzed and prospect students for a particular course could be identified with minimum dropout rate [368]. Data Mining using student behavioral data has been discovered the difference between successful/unsuccessful students in terms of level of participation in various activities.

10.8.5 Data Mining for Knowledge Modeling

It is designed on the basis of the interactions between students and learning system. It is interesting to obtain a correlation between correctness of students and the amount of time they spent to practice it before answering [369]. This data model assists the teachers to discriminate between students who are trying and who are not. The discovered knowledge then helps the teachers to use different instructional polices for each group of students.

10.8.6 Data Mining for Student Behavior Modeling

It is used to predict the behavioral outcome of the student by measuring how much time a student has spent online, whether it has completed a defined course, changes the documents in classroom or school, attendance rate, tardiness ratio etc. It also provides hints to the teachers so that they can easily understand the learning capacity of particular students. By using similar manner the shifting of classroom, assigning a substitute teacher as well as lack of attention of a teacher for an online learning system can be also predicted [370].

10.8.7 Data Mining Based Learning Model for Teachers

A Data Mining model can observe the detailed activities of teachers, instructors, administrators and policy makers. This Data Mining based model can also be used to improve efficiency, relevance and effectiveness of the teachers particularly in a higher educational institution [371]. The performance of a teacher and the success of a student are considered to calculate the amount of pedagogical support that a student needs or has received. The discovered knowledge also indicates how the pedagogical support is affecting the learning resources.

10.8.8 Data Mining for Placement Profiling

Data Mining based modeling are often used to categorize the profile of students based on their personal learning data, responsiveness and demographic data for proper placement suggestion [372]. This information would be taken further to classify the fresher for long-term achievements.

10.8.9 Data Mining for Course Suggestion Model

Proper course modeling in Education Data Mining is performed on the basis of associated skill of the student, previous success rate of students and latest advancement in relevant field [373].

10.9 Data Mining Application in Telecommunication Industry

One of the important tasks of Data Mining in telecommunication industry is to utilize the network capacity. The network capacity utilization is essential for telecommunication market expansion and also for policy decision regarding customer services [374].

With the help of Data mining, it is also possible to identify the underlying patterns and structures of most popular services preferred by customer groups. Such type of analysis can optimize the investments in network facilities for better customer services and costly expansions could also be planned accordingly. Using Clustering approach, companies can separate the customers into different segments based on their characteristics and take different actions to increase revenue and also avoid losses on the basis of their usage patterns. Predictive Data Mining is used to classify the customers as well as their likelihood can also be estimated in advance.

10.9.1 Credit Card Fraud Detection

The main attributes of credit card fraud data are account holder details, transaction limits, previous purchase habits etc. By using clustering, any transaction data can be grouped as safe transaction with very high, high, low and no risk categories. In this manner it is possible to predict the most frequent usage pattern and any possible misuse of the particular card. Figure 10.9

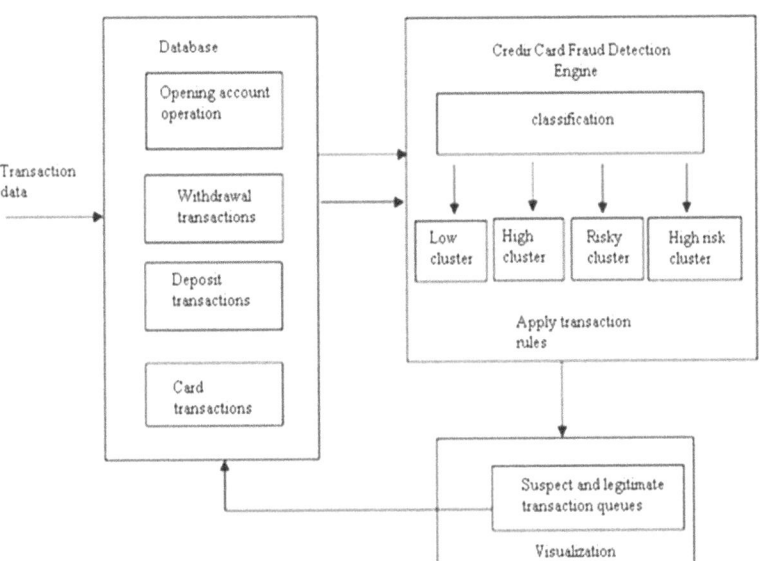

Figure 10.9 Working Model for Credit Card Fraud Detection

describe the working model for credit card fraud detection. Various categories of transaction data is required to create a central repository. A credit card fraud detection engine refers that repository and generates different cluster on the basis of degree of risk involved. The characteristics of individual cluster can be visualized with various class labels to discriminate safe and suspicious transaction.

10.10 Conclusion

Most of the organizations such as Health, Finance, Business, Insurance and Telecommunication are directly dealing with customers and generating vast data. All these organization have common requirements such as customer satisfaction, resource utilization, intelligent policy plans, fraud detection and enhanced business outcomes. Data Mining as universal strategic management tool is a proven solution for their latest demand because of competitive environment. In this present knowledge driven society Data Mining is an upcoming solution to address current demands of complex business organizations where time and money both are the leading concern for everyone.

11

E Governance Data Mining- A SWOT and PESTLE Analysis

11.1 Introduction

Presently all real time systems are using information and communication technologies and generating huge data which is the key asset of any organization. Extracting valuable information from huge data repository is a crucial task and certain well defined processes are required to be followed in such type of knowledge discovery phases. Data Mining is a way to discover knowledge form databases for efficient decision making. Data mining is gaining popularity in day by day from health care to law enforcement and the necessity of efficient data mining tool is clearly established.

Figure 11.1 illustrates various phases of data mining which are complex in nature. It needs data sources for continuous collection of data items. The phase extract, transform and load are important to prepare the data for mining purpose. This phase basically cleans the data and transforms it into universally acceptable format. Data preprocessing requirements are different for different set of data generated by heterogeneous data sources. Identification appropriate data preprocessing technique is an important task which needs knowledge of the domain as well as data expertise. Establishment of a data warehouse is also challenging due to its hardware/ software constraints. A data center could be utilized for the purpose of efficient data storage and used to have a central repository which is the key requirement of data warehousing. Cloud computing is an upcoming technique provides solution for data center establishment but still in early stages of development.

Data mining techniques such as clustering and classification algorithms are used for manage and retrieve information as per need. Association rule mining techniques could be applied for discovering the interrelationship between two or more objects. It is also used for establishing relationships between different set of events. The aim of data mining techniques is to extract knowledge and

E Governance Data Center, Data Warehousing and Data Mining: Vision to Realities, 237–248.

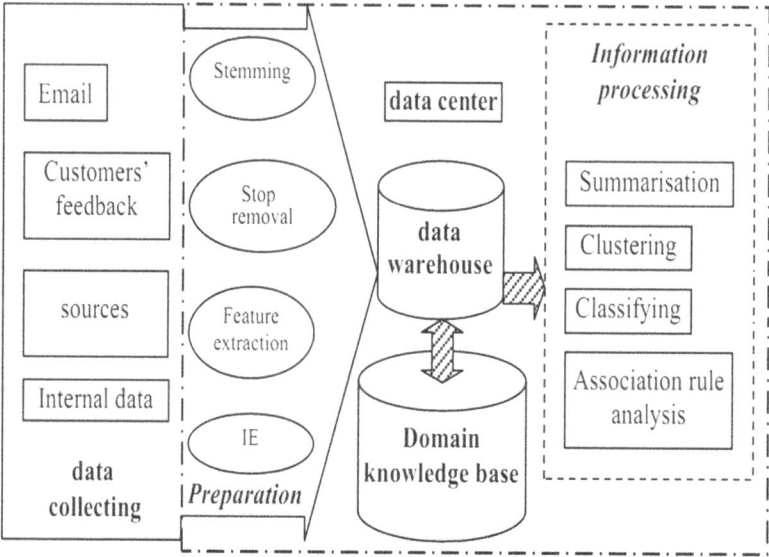

Figure 11.1 Phases of Data Mining

represents it to the decision maker in a more concise format. Both textual and visual summarization methods could be utilized.

E Governance services in worldwide have been enormously admired since last few years because it has very large target segments of citizens, governments and business parties. E Governance is certainly a revolution in the government services, to provide full availability, accessibility, efficiency with transparency and accountability to the whole society. The most essential requirement of E Governance projects are computerization of different government organizations. Increasing trend of computerization in different government departments such as regulatory, developmental including social welfare involve a large data.

In any government system decisions are always based on previous practices and experiences. Once Data Mining system has been adopted, conventional practices have been improved and best practices could be introduced to complete the task in time. Moreover Data Mining could facilitate quick and useful knowledge extraction from huge data sources at reduced costs and finally ever-increasing organization growth and development opportunities. So, it is clearly establish that present E Governance system should well equipped with data mining tools for efficient decision making and it is also needed to identify the challenges associated in this integrated approach which utilizes Data Centers, Data Warehousing and Data Mining for good governance. Two popular analysis techniques, SWOT and PESTLE analysis are presented here to justify the role of Data Center, Data Warehouse and Data Mining for successful implementation of E Governance in India as well as across the world.

11.2 SWOT Analysis

SWOT analysis is a way to investigate any project or areas under four parameters i.e. S-Strengths, W-Weakness, O-Opportunities and T-Threats [375] [376]. Here an integrated approach of E Governance has been considered which consists Data Center, Data Warehousing and Data Mining techniques for efficient and transparent government decision making. A SWOT analysis for all these emerging techniques has been presented to highlight the challenges, priorities, benefits and upcoming trends.

11.2.1 SWOT Analysis Data Center

11.2.1.1 Strengths:
- Data centers provide a secure environment.
- Data centers have different kind of connections such as GSM/CDMA. These is sufficient to fulfill all stakeholder requirements.
- Data center facilitate hardware, networking, application support for any E Governance system.
- It also facilitates security and backup for all E Governance operations.
- The data center needs trained manpower so skills development could be done for efficient management of data center.
- Data Centers provides full power backup and helpful to maintain round the clock availability of G2C interface.

11.2.1.2 Weaknesses:
- Establishment of data center needs high development, operational and maintenance cost.
- Deficiency of staff for data centers is another weakness. It can be removed by hiring IT professionals who are experienced in IT infrastructure management services
- Non –commitment of staff also creates problem. It could be solved by allocating task on target basis in data centers.

- Policies should be framed for data center business.

11.2.1.3 Opportunities:

- Since each and every government organization is automated now and serving online so data center could be utilized fully in all district/ states.
- Data center works as government units so it can easily protect government data from unauthorized access.
- We can establish data centers on present network so it reduces setup cost for government organization.
- Servers established at data center also provided to business units at reasonable rate.
- Data centers are essential part for distributed computing environment which is best suited for current working environment of government organizations.

11.2.1.4 Potential Threats:

- Cyber threats such as hacking, Denial of service attacks, spoofing, phishing etc.
- Environmental threats such as Earthquake, Tsunami, Flood etc.
- Retaining of skilled Manpower.
- Technology changes.

11.2.2 SWOT Analysis of E Governance

11.2.2.1 Strengths:

- E governance simplifies work routines and provides quality service.
- E Governances is adapted everywhere in the world and experiencing several advantages.
- Reduced number of personnel required because computerization minimizes the need of labor.
- More citizens' participation could be achieved.
- Provide more transparency to administrative work.
- Assisting in establishing bi-lateral and global relations among people and nations.

11.2.2.2 Weaknesses:

- Literacy and computer literacy are one of the most challenging issues because of digital divide in less developed countries.

- Information technology is a fast growing field so there is a considerable gap has been identified between the rate of technology change and social and cultural development.
- Different nations are having different GDPs thus their expectation are also different regarding E Governance projects outcomes.

11.2.2.3 Opportunities:

- Political, social, cultural acceptance.
- Information Technology solution are becoming cheaper.
- Various modes of Internet connectivity options are available to cater varying needs of stakeholders.
- Promotes citizen participation and best suited for democratic nations.
- Decrease the corruption opportunities and facilitate transparent government.
- Sophisticated working environment is available to accommodate the manpower.
- Highly interactive G2C interfaces are available to fulfill citizen needs.
- Create a citizen-centric government framework.

11.2.2.4 Threats:

- Digital divide problem may be prominently experienced after widely acceptance of E Governance projects.
- Higher infrastructure cost.
- Cyber crimes such as hacking, Denial of service attacks, spoofing, phishing etc.
- IT expert are required for smooth functioning of E Governance system.
- Security and copyright related issues.
- Social, Legal and ethical concerns
- Dependency on technology, If technology changes, high cost project maintenance is required.

11.2.3 SWOT Analysis of Data Mining

11.2.3.1 Strengths:

- Data Mining helps administrator to explore knowledge from huge dataset by using clustering, classification, regression, association techniques.
- By using data mining techniques it is possible to learn from historical datasets which in turn develop better understanding about a situation and better strategy could be developed.

- For data mining a data warehouse is an essential component which stores data in a central place. This central data repository eliminates the possibility of data redundancy and it also ensures transparent government decision making for various schemes such as scholarships etc.

11.2.3.2 Weakness:

- Data Mining needs huge data about any domain where it is subjected to apply. There is always risk of breach of personal privacy and the data may be misused in unethical manner by the intruders.
- The Data Mining is used for exploring useful knowledge from huge data but how this information may be utilized further is a critical issue. Hackers may use this valuable information for their personal benefits.
- The performance of Data Mining may suffer from false positive, false negative and Misclassification which may not be accurate for further policy making. If such information is used for strategic planning a severe consequences may be possible.

11.2.3.3 Opportunities:

- **User's satisfaction:** Various Data mining techniques are used to identifying the requirements of the users such as customers, IT professionals, Business Analyst, Doctors etc. Data mining is also helpful to analyze organizational data for efficient decision making.

- **Increasing resource utilization:** Data mining explores hidden patterns, clustering and association from huge datasets. These techniques are useful to find out correlation between resources available and resources utilized. In this manner efficient resource utilization could be possible on the basis of hidden patterns, association and clustered identified after applying data mining techniques.

- **Close observation of citizen/stakeholder:** Data mining could be applied with different person profiles to check their day-to-day activities. Any abnormal behavior of a person could be easily identified as an outlier and the same may be correlated in terms of abnormal activity of that person. In this manner person profiling could be done to avoid any fraudulent as well as suspicious activities related to banking system, income tax department and homeland security.

- **Efficient policy making for E Governance:** Clustering is used to find out similar patterns that may be in terms of social, environmental and political similarities. All regions belonging to same cluster may be treated under one policy and in this manner policy decision may be taken quickly with less effort.

11.2.3.4 Threats:

- All Data Mining Solution providers are facing cut-throat competition. They all are struggling for the market share for their products.
- Application of Data Mining in any organization for the specific need of that organization may raise the cost of the product and it becomes challenging for small organizations. Data used in Data Mining may face the risk of privacy threat.

11.3 PESTLE Analysis

PESTLE analysis is done on the basis of 6 parameters **P**-Political, **E**-Economic, **S**-Social, **T**-Technological, **E**-Environmental and **L**-Legal and describes effect of environment factors on strategic management component [377]. It is a kind of external analysis which should be considered at the time of research. It has several environmental parameters which should be considered for understanding the circumstances and for deciding directions for various operations [378] [379].

11.3.1 PESTLE Analysis for Data Mining

- **Political Factors**: In data mining, political factors are mainly the influence of political institution on the projects which are under implementation. It primarily includes government policies regarding usage, funding, grants and initiatives of data mining application in any regular E Governance projects.
- **Economic Factors**: The primary and foremost economical factor is the cost of data warehouse and data center which are the key component of a data mining operation.
- **Social Factors**: Data mining is becoming popular to a common man. Various social welfare issues such as education, health, agriculture, rural development are the most suitable application areas of data mining. Some recognized examples are disease prediction, e-learning management, student performance prediction and weather forecasting system.
- **Technological Factor**: Data mining software uses latest techniques available in the market. These technologies will help in collecting data, analyzing data and developing knowledge. All data mining solution are costly

but recent development in open source software showing the opportunity of low cost data mining solutions suitable for E Governance applications.

- **Legal Factor**: From a legal viewpoint, the main problem in data mining is data protection. It is important to ensure proper privacy protection in data mining applications so that data could not be misused by anyone for their personal benefits.
- **Environmental Factor**: It is a software based approach. So, there is no harm to the environment.

11.3.2 PESTLE Analysis for Data Center

- **Political Factor**: Data centers are emerging technologies and currently powered by cloud computing features. It needs high budget, skilled manpower, and suitable place for establishment. All these requirements may face political influences.
- **Economic Factor**: Data centers are working either with government organization or with business organization. Most of the business organization are capable enough to bear the high budget of data center establishment but since the government project are already a limited budget setup so bearing the expenditure for data center became a challenging task. This could be overcome only when if the same data center could be used to facilitate small size enterprises for their needs on cost sharing basis. In this way the overall cost could be shared and government organization can also maintain their dedicated data centers.
- **Social Factor**: In present situation the popularity of web based application, social networking websites, e-learning application are gaining momentum. All these application involves huge data generation so the need of data center is clearly established.
- **Technological Factor**: Different type of techniques can be combined with data centers such as web server or file server, email etc. These techniques provide more computing power to data centers. These techniques can be used whenever they need.
- **Legal Factor**: In data center legal issues are related to service agreement, guarantee, warrantee associated with various products under use. If party fails to follow the agreement condition, the data center authorities can take an action against him.
- **Environmental Factor**: Data center can work on greener alternatives of latest IT technologies which will help in reducing energy consumptions.

11.3.3 PESTLE Analysis of E Governance

- **Political Factors**: E Governance is a way of transformation of conventional government practices by means of Information and communication technologies. E Governance improves various existing processes using online applications and provides transparency, accountability and faster delivery to the stakeholders. Such type of changes in existing practices may face resistance due to some political issues. Although the advantages achieved by this may be appreciable by all but changing the existing work practices at the beginning is difficult and requires higher degree of encouragement.

- **Economic Factors**: For successful implementation of E Governance computer literacy, bandwidth, processing power, memory and automation is required. All such arrangement needs initial investments with proper planning and budgeting. This indicates that E Governance projects should be highly motivated towards their end objectives and their success must be guaranteed to justify the effort involved.

- **Social Factors**: Successful implementation of E Governance is now changing the shape of the society form various dimensions. Emergence of online E Governance services changes the situation from in-line to on-line for citizens. Quicker delivery of services created a smooth working environment for government officials as well as citizens and enhances the overall job satisfaction for both. Popularity of internet increases women participation in latest technological trails and bringing them to the mainstream of the society. All these factors are collectively contributing towards prosperous society with good quality of life.

- **Technological Factors:** E Governance reforms need technological advancement in its all phases of implementation. This includes latest hardware, software, networking, and skilled manpower requirements. The technological requirements are hard to fulfill because of higher cost and fast changing trends. Skill development for any particular technology is also a difficult task and becomes challenging when it has to match the requirement of fast changing technology scenario. All such issues indicate that there is a need of outstanding plan to maintain E Governance projects successfully for longer period of time.

- **Legal Factors:** Online applications involve various legal issues such as privacy protection, authenticity, authorization which are difficult to maintain. Automation of various E Governance project also needs variety of system software, application software with their usage license which is a costly affair.

Table 11.1 Comparison Between SWOT and PESTLE Analysis of E-Governance

	Strength(S)	**Weakness(W)**	**Opportunity(O)**	**Threat(T)**
Political(P)	Public policy (e.g. cata center License Scheme) Cooperation between the public and private sectors	Conservation in trying e-Serviees	Political wilingness	Cyber terrorism and cyber crimes Security breach and copyright issue
Economical(E)	Economic policies (e.g. Data Quality Class) Funds for e-Govt. services to improve social and physical infrastructure Low cost of Internet subscription	Unemployment Poverty Digital divide	IT-profcient people can have better opportunity for employment	Higher cost of living and higher Broadband Subscription.
Social (5)	Educational system (e.g. national IT Literacy program) Recruitment of foreign talents Tech-savvy population (e.g. e-district, resource Re-use scheme)	workers and older generation are computer illiterate	Encourage citizen participation in online shopping, e-Services etc.	The rapid development of mobile and SMS Technology
Technological(T)	innovation Availability of Powerful tools and techniques. Utility software Open source software	Some government websites are unfriendly-user Over-capacity of the internet highway due to heavy traffic	Broadband facilitates faster connection.	Dependency on IT, i.e. small technical Problems disrupt the entire networks.
Legai(L)	Provices authorize users and licensed products to the end users	Costly and short term agreement	Emergence of GNU general pubic licenses	Piracy, hacking, phishing
Environmental(E)	More citizen participation including women, increasing number of online application, reduced waiting time, faster delivery and creating good working environment	Computer Illiteracy, unreachability, digital divide	Green ICT	Power dissipation, global warming, e waste

- **Environmental Factors:** Adoption of E Governance environment in present era is changing the way of life. It has enhanced the communication speed and minimizes a lot of paper work. It also reduced the waiting time involved in various government processes and subsequently created a healthy environment for our developing society.

11.4 Comparative Study of SWOT and PESTLE Analysis of Data Center for E Governance

The Comparative Study of SWOT and PESTLE Analysis of Data Center for E-Governance is indicated in the Table 11.1.

11.5 Conclusion

In this chapter SWOT and PESTLE analysis has been done to identify critical parameters of the E Governance Data Mining. E Governance Data Mining through Data Center is an initiative to integrate Government, Citizen and Business Organization for a dignified goal of setting good governance. It is an attempt to learn from past practices and discover unexpected facts from huge data stored in Data Warehouses using Data Centers. It helps to get better citizen participation in various government operations to facilitate greater efficiency. It is clearly established that successful implementation of E Governance needs persistent management, seamless coordination, consistency, standardization and interoperability. There is also a need of a legal framework which is used to regulate the E Governance system. The faith on E Governance could be strongly developed if availability, accessibility, security and privacy could be ensured. Although there are high early investments but manifold benefits could be experienced if required skill is available and there are fair policies for knowledge sharing.

References

1. Jafari, S.M., Ali N.A., Sambasivan M., Said, M.F., "A respecification and extension of DeLone and McLean model of IS success in the citizen-centric e-governance," *Information Reuse and Integration (IRI), 2011 IEEE International Conference on*, vol., no., pp.342–346, 3–5 Aug. 2011

2. Hossain, A.L., "A study on computerized Emigration Clearance system of Bangladesh: Step toward e-Governance," Computer and information technology, 2007. iccit 2007. 10th international conference on , vol., no., pp.1–6, 27–29 Dec. 2007

3. Uwano, H., Kamei, Y., Monden, A., Matsumoto, K., "An Analysis of Cost-Overrun Projects Using Financial Data and Software Metrics," Software Measurement, 2011 Joint Conference of the 21st Int'l Workshop on and 6th Int'l Conference on Software Process and Product Measurement (IWSM-MENSURA) , vol., no., pp.227–232, 3–4 Nov. 2011

4. Mahmud, K., Gope, K., "Prospects of Implementing Short Message Service (SMS) Based E-government Model in Bangladesh," Computer Technology and Development, 2009. ICCTD '09. International Conference on , vol.1, no., pp.153–157, 13–15 Nov. 2009

5. Official Website of e-Government for Development (n.d.), Retrieved on November 2012 from http://www.egov4dev.org/transparency/

6. Official Website of Warwick (n.d.), Retrieved on November 2012 from http://www2.warwick.ac.uk/knowledge/business/thinklocal/

7. Official Website of e-government databases (n.d.)," Indian Journal of Computer Science" Retrieved on November 2012 from www.ijcse.com/docs/IJCSE10–01-02–01.pdf

8. SokChuob, Pokharel, M., Jong Sou Park , "The future data center for e-governance," Advanced Communication Technology (ICACT), 2010 The 12th International Conference on , vol.1, no., pp.203–207, 7–10 Feb. 2010

9. Official Website of CDAC (n.d.) , Retrieved on November 2012 from http://pune.cdac.in/html/dwh/dwhegov.aspx

10. Official Website Data Mining tools (n.d.) , Retrieved on November 2012 from http://www.dataminingtools.net/wiki/applications_of_data_mining.php

11. Official Website Waikato Environment for Knowledge Analysis(n.d.), Retrieved on November 2012 fromhttp://www.excelsior-usa.com/jetcs00002.html

12. Jiawei Han, Micheline Kamber (2006) "Data Mining: Concepts and Techniques, Third Edition (The Morgan Kaufmann Series in Data Management Systems)", Publisher: Morgan Kaufmann 2006.

13. "History of Data Mining"(n.d.) Retrieved on July 2010, from www.data-mining-software.com/data_mining_history.htm.

14. Buchanan, B.G.,(n.d.) "Brief History of Artificial Intelligence", Retrieved on "January 2010 from http://www.aaai.org/AITopics/bbhist.html.

15. Kantardzic, Mehmed (2003) "Data Mining: Concepts, Models, Methods, and Algorithms". John Wiley & Sons. ISBN 0471228524. OCLC 50055336.

16. Dunham, M.H. (2003), "Data Mining introductory and advanced topics" Upper Saddle River, NJ: Pearson Education, Inc.

17. Chaudhuri Surjit, Dayal Umeshwar, (1997), "An Overview of Data Warehousing and OLAP Technology", Retrieved on November 2010, from www.c Ralph Kimball (1996) The Data Warehousing Tool Kit , John Willy & Sons 1996.

18. Ralph Kimball (1996) "The Data Warehousing Tool Kit" , John Willy & Sons 1996.

19. Jeffrey A. Hoffer, et al., (2005) "Chapter 11:Data Warehousing" in Modern Database Management 7th Edition 2005 by Prentice Hall.

20. Bellinger, G., Castro, D. & Mills, A. (1999) "Data, information, knowledge and wisdom". http://www.outsights.com/systems/dikw/dikw.htm.

21. Liebowitz, J. (2000) "Building Organizational Intelligence: A Knowledge Management Primer". Boca Raton, CA: CRC Press. Knowledge Acquisition" (n.d.) Retrieved December 2005 from www.epistemics.co.uk/Notes/63–0-0.htm.

22. Knowledge Acquisition" (n.d.) Retrieved December 2005 from www.epistemics.co.uk/Notes/63–0-0.htm.

23. Burk Mike (n.d.) "Knowledge Management: Everyone Benefits by Sharing Information" Retrieved November 2005 from www.tfhrc.gov/pubrds/novdec99/km.htm.

24. Riley Thomas B.(2003) International Tracking Survey Report '03 Number Two "Knowledge Management and Technology" http://www.rileyis.com/publications/research_papers/tracking03/IntlTrackingRpt June03 no2.pdf.

25. Wright Peggy (1998) "Knowledge Discovery in Databases: Tools and Techniques" www.acm.org/crossroads/xrds5–2/kdd.html.

26. Schroder, H (2004) "Marrying Knowledge Discovery in databases (KDD) with technology intelligence (TI) - avenue to paradise or blind alley?." Engineering Management Conference, 2004. IEEE International Volume 1, Issue , 18–21 Oct. 2004 Page(s): 276 - 282 Vol.1..

27. Fayyad, U.M, Piatetsky-Shapiro, G., & Smyth, P. (1996). "From Data Mining to Knowledge Discovery in Databases". AI Magazine, 173, pp. 37–54.

28. Ruey-Chyi Wu, Ruey-Shun Chen, Chen, C ,(2005) "Data Mining application in customer relationship management of credit card business". Computer Software and Applications Conference, 2005. COMPSAC 2005. 29th Annual International Volume 2, Issue , 26–28 July 2005 Page(s): 39 - 40 Vol.

29. Official Website of CRISP-DM 1.0 (n.d.) "Step-by-Step Data Mining Guide", Retrieved on November 2010 from http://www.crisp-dm.org/CRISPWP-0800.pdf.

30. Kurt Thearling,(n.d.) "An Introduction to Data Mining" Retrieved December 2005, from www.thearling.com/text/dmwhite/dmwhite.htm.

31. Goharian and Grossman, (2003) "Data Mining Classification", Illinois Institute of Technology, http://ir.iit.edu/~nazli/cs422/CS422-Slides/DM-Classification.pdf.

32. Apte C. & Weiss S.M. (1997) "Data Mining with Decision Trees and DecisionRules" T.J. Watson Research Center http://www.research.ibm.com/dar/papers/pdf/fgcsaptewe iss _with_cover.pdf.

33. Jovanovich N, Milutinovic V, Obradovic Z, (2002) "Foundations of predictive Data Mining Neural Network Applications in Electrical Engineering", 6th Seminar on Volume , Issue , 2002 Page(s): 53 – 58.

34. Bezdek, J., (1981) "Pattern Recognition With Fuzzy Objective Function Algorithms", Plenum. New York.

35. Hoppner, F., Klawonn, F., Kruse, R., and Runkler, T., (1999) "Fuzzy Cluster Analysis Methods for Classification, Data Analysis and Image Recognition", John Wiley and Sons Ltd.

36. Vapnik, V. (1998). "The support vector method of function estimation".

37. Cristianini, N., & Shawe-Taylor, J. (2000) An Introduction to Support Vector Machines. Cambridge University Press.

38. Dietterich, T. G., & Bakiri, G. (1995) "Solving multiclass learning problems via error-correcting output codes". Journal of Artificial Intelligence Research, 2, 263–286.

39. N.Chistianini and J. Shawe-Taylor, (2000) "An Introduction to Support Vector Machines, and other kernel-based learning methods", Cambridge University Press, 2000.

40. Dawn P. Gill et.al., (2009) "Comparison of Regression Models for the Analysis of Fall Risk Factors in Older Veterans", Vol. 19, Issue 8, Pages 523–530.

41. Fox, J. (1997). Applied Regression Analysis, Linear Models, and Related Methods (p. xxi, 597 p.).

42. Andrew Gelman, Jennifer Hill (2007) "Data analysis using Regression and multilevel hierarchical models" Cambridge ; New York : Cambridge University Press, 2007.

43. Jain A.K, Murty M.N., Flynn P.J., (1999) "Data Clustering: A Review" ACM Computing Surveys, 31, 3:264–323.

44. Leonardo E. Auslender, (n.d.) "On Analytical Tools for Market Basket (Associations) Analysis". SAS Institute, Research & Development http://nymetro.chapter.informs.org/prac_cor_pubs/Ausleder-On-market-basketanalysis-May-04.pdf.[62].

45. J. Liu, et. al., (2006) "Distance-Based Clustering of CGH Data". Bioinformatics, 22(16):1971–1978.

46. A.K. Jain , M.N. Murty, P.J.Flynn (1996) "Data Clustering: a review" ACM Compute, Surveys 31,1996.

47. Hamerly, G. and Elkan, C., (2003), "Learning the K in K-means", in Proceedings ofthe 17th Annual Conference on Neural Information Processing Systems, British Columbia, Canada.

48. Richard Heeks (2001) "Understanding E Governance for Development" Institute for Development Policy and Management, University of Manchester, UK Published by: Institute for Development Policy and Management, ISBN: 1 902518934 http://www.iimahd.ernet.in/egov/ifip/dec2001/article3.htm.

49. Kashif Farooq, et al. (2006) "Devolution of e-Governance among Multilevel Government Structure" University of Management Sciences (LUMS), Pakistan 1–4244-0674-9/06/$20.00 ©2006 IEEE.

50. Gartner Group, (2000) "Key Issues in E-Government Strategy and Management," Research Notes, Key Issues, Retrieved on February 2010.

51. Garnter Group, (2000) "Gartner's Four Phases of E-Government Model, Research Notes, Key Issues, Retrieved on February 2010.

52. Christopher Baum and Andrea Di Maio, (2000) "Gartner's Four Phases of EGovernment Model", Retrieved on February 2010.

53. http://twocircles.net/data_bank/telephone_subscribers.html "Retrieved in March 2010.

54. Government of India second administrative reforms commission 11th report "promoting e-governance the smart way forward" status report http://arc.gov.in/11threp/ARC_11thReport_Ch3.pdf.

55. http://www.wikiprogress.org/index.php/E-Government_Readiness_Index

56. United Nations, E-Government for the People, E-Government Survey, 2012

57. Framework for a Set of E-Government Core Indicators, ITU, 2011

58. GSM Association (2011). Africa Now the World's Second Largest Mobile Market, Reports GSMA. GSM World, 9 November 2011. Available from http://www.gsma.com/articles/africa-now-the-world-s-second-largest mobile-market-reports-gsma/20866. Accessed January 2012

59. Janet Kaaya, "Implementing e-Government Services in East Africa: Assessing Status through Content Analysis of Government Websites", Electronic Journal of e-Government Volume 2 Issue 1.

60. Arjan de Jager and Victor van Reijswoud, "E-Governance - The Case of DistrictNet in Uganda", Thematic Networks

61. http://zambia.co.zm/

62. www.unza.zm/zamlii/

63. http://www.mozambique.mz/

64. Macueve, G. (2008) e-Government for Development: A Case Study from Mozambique, The African Journal of Information Systems, Volume 1, Issue 1, pp. 1–17

65. www.ethiopia.gov.et/

66. http://niledialogue.org/discus/messages/13/4266.html?1005156670

67. http://mail.kenya.go.ke/

68. www.gta.gov.zw/

69. www.malawi.gov.mw/

70. www.republique-djibouti.com/23.htm

71. www.tanzania.go.tz/

72. Wilfred Uronu Lameck, "The role of e-governance in facilitating information needs in higher learning institutions: The case of Mzumbe University in Morogoro, Tanzania", Journal of Public Administration and Policy Research Vol. 36, pp. 184–187, June 2011

73. Mark Burke, "A Decade of e-Government Research in Africa", LINK Centre, Faculty of Humanities, University of the Witwatersrand.

74. http://www.spm.gov.cm/index.php?L=1

75. Statistical, Economic and Social Research and Training Centre for Islamic Countries (SESRIC), "E-Government Development and E-Participation, The Performance of the OIC Member Countries"

76. Muhammad Saadi, Alhusein Almahjoub, "E-Governance in Libya – Where we are and Where to Go"

77. http://www.tunisie.gov.tn/index.php?lang=french

78. www.pm.gov.tn/pm/content/index.php?lang=en

79. http://www.egypt.gov.eg/english/

80. http://www.sd.undp.org/projects/tokten%20work.htm

81. http://www.sd.undp.org/projects/tokten.htm

82. http://www.sd.undp.org/projects/tokten%20req.htm

83. http://www.sd.undp.org/projects/tokten%20for%20institutions.htm

84. Bruce Mazengera, "Factors Contributing To Successful E-Government Implementation in Southern African Development Community (SADC) Countries", Innovation and Knowledge Management: A Global Competitive Advantage.

85. http://www.gov.za/

86. http://www.eisa.org.za/WEP/sou4.htm

87. http://www.gov.ls/home%5Cdefault.php

88. www.gov.sz/

89. www.grnnet.gov.na/

90. http://www.gov.na/

91. "Botswana's National e-Government Strategy, 2011–2016", Republic of Botswana

92. "E-governance and Citizen Participation in West Africa: Challenges and Opportunities", The Panos Institute West Africa & The United Nations Development Program, 2012

93. http://www.ghana.gov.gh/index.php/news/features/11367-ghana-to-commence-e-government-project

94. http://www.ghanaceg.org/programmes.html

95. Denise Clarke, "E -Governance in Ghana: National Information Clearing House", Capacity Development Officer, IICD

96. Dr. Fatile, Jacob Olufemi, "Electronic Governance: Myth or Opportunity for Nigerian Public Administration", International Journal of Academic Research in Business and Social Sciences, September 2012, Vol. 2, No. 9

97. http://www.thenationonlineng.net/2011/index.php/saturday-magazine/travels-on-saturday/52135-world-%E2%80%98n-traveland-unveils-self-service-kiosk-in-nigeria.html

98. http://www.wangonet.org/projects/index.php

99. "Senegal: Strengthening Capacities for Monitoring and Evaluation of Good Governance", GAP initiative, http://www.gaportal.org/undp-supported/senegal

100. Pierre Dandjinou, "Assessing e-Governance Impact Country case studies: Cape Verde and Senegal", e-Governance Advisor committee, UNDP

101. Yong, James S.L. (2004). Promoting Citizen-Centered Approaches to E-Government Programmes – Strategies and Perspectives from Asian Economies. Paper presented at the Second APEC High-Level Symposium on E-Government, Acapulco, Mexico. (6–8 October 2004)

102. http://www.bahamas.gov.bs/wps/portal/public/gov/government/eServices/!ut/p/b0/04_Sj9CPykssy0xPLMnMz0vMAfGjzOKN3f19A51NLHwtAhxdDTwNQ_z9Ag19DP2djPULsh0VAZl2VXA!/

103. http://cceeg.dec.uwi.edu/drupal/?q=bahamas

104. http://www.mes.gov.bb/pageselect.cfm?page=89

105. http://www.carib-is.net/integrated-justice-information-system-ijis

106. http://www.asycuda.org/dispcountry.asp?name=Barbados

107. http://www.ict-pulse.com/2012/05/snapshot-e-government-update-2012/

108. Abraham Sotelo-Nava, "e-Government in Mexico Participation and Inclusion", United Nations, Head of e-Government and IT Policy, Ministry of Public Administration, Mexico

109. http://www1.american.edu/initeb/cs6223a/egovernment.htm

110. Wolff, L. and C. Castro (2003). Education and Training: the Task Ahead. After the Washington Consensus: Restarting Growth and Reform in Latin America. P. Kuczynski and J.Williamson, Eds., Washington, D.C.: Institute for International Economics

111. Mena, S. R. "Ecuador's Experience in E-Governance." Encyclopedia of Digital Government. IGI Global, 2007. 437–441. Web. 15 Mar. 2013

112. "From Municipalities in Chile and Peru", United Nations Publication, LC/W.31, October 2005

113. Paulo Alcântara Saraiva Leão, "Electronic Government in Brazil", Global e- Government for Development Conference - University of Maryland – USA, 2006

114. http://www.thisischile.cl/Article.aspx?id=4810&sec=419&idioma=2&eje=

115. Roy, Jeffrey and Longford, John (2008). Integrating Service Delivery across Levels of Government: Case Studies of Canada and Other Countries, IBM Center for the Business of Government

116. http://www.mgs.gov.on.ca/en/IAndIT/STEL02_046918.html

117. Barbara Ann Allen, Luc Juillet, Gilles Paquet, Jeffrey Roy, "E-Governance & Government Online in Canada: Partnerships, People & Prospects",

118. "E-Government Strategy", President's Management Agenda for E-Government, USA, 2002

119. Codrin-Marius Teiu, "An overview over the worldwide development of e-government", Munich Personal RePEc Archive, http://mpra.ub.uni-muenchen.de/36470/

120. Adegboyega Ojo, Mohamed Shareef and Tomasz Janowski, "Electronic Governance in Asia: State of Play, Impact and Bridging Internal Divide", Center for Electronic Governance, United Nations University

121. http://www.egov.kz/wps/portal/index

122. Kassen, Maxat, "E-Government in Kazakhstan: Realization and Prospects" (2010). 2010. Paper 6

123. Sandjar Saidkhodjaev, "e-Governance portfolio: UNDP Uzbekistan", UNDP Global Community of Practice meeting on e-Governance and Access to Information, 2009

124. http://www.undp.tj/site/index.php/en/our-programme/good-governance/61-bomnaf

125. www.bomca.eu/en/tajikistan.html

126. Adegboyega Ojo, Mohamed Shareef and Tomasz Janowski. Electronic Governance in Asia - State of Play, Impact and Bridging Internal Divide. The Internet and Civil Society in Asia, Hermès, 55, 2009.

127. http://www.lgcns.com/service/reference/view/107/korean-intellectual-property-office-network-kiponet-system

128. http://www.hometax.go.kr/index.jsp

129. http://www.pps.go.kr/user.tdf?a=common.HtmlApp&c=3001&page=/english/what_we_do/geps/overview/overview.html&mc=PE_04_16_01

130. http://www.mopas.go.kr/gpms/view/english/national/national01.jsp

131. http://www.e-gov.go.jp/doc/e-government.html

132. http://jcmc.indiana.edu/vol9/issue4/zhou.html

133. Ross, N., L. Hutton and L. Peng (2004). Revolutionary E-Government Strategies across Asia-Paci? c – Strategy White Paper. Alcatel Telecommunication Review (3rd Quarter)

134. http://www.nadra.gov.pk/index.php/about-us

135. http://www.e-government.gov.pk/gop/index.php?q=aHR0cDovLzE5Mi4xNjguNzAuMTM2OjkwODgvZWdkc2l0ZTA1Lw%3D%3D&hl=2ed

136. http://textus.diplomacy.edu/thina/txgetxdoc.asp?IDconv=3249

137. Dmitry Pozhidaev, "Local E-Governance in Afghanistan: Dream or. . . ?", UNDP E-Governance CoP Meeting, 2009.

138. http://www.egov.iist.unu.edu/cegov/projects/EGOV.AF-e-Government-in-Afghanistan

139. http://mcit.gov.af/en/page/7081

140. http://www.egov.vic.gov.au/focus-on-countries/asia/sri-lanka/e-government-sri-lanka-archive.html

141. http://www.icta.lk/en/e-sri-lanka.html

142. http://nwb.bcc.net.bd/index.php?option=com_content&view=article&id=56&Itemid=89&lang=en

143. Yong, James S.L. (2004). Promoting Citizen-Centered Approaches to E-Government Programmes – Strategies and Perspectives from Asian Economies. Paper presented at the Second APEC High-Level Symposium on E-Government, Acapulco, Mexico. (6–8 October 2004)

144. http://www.oasissingapore.com/

145. http://www.egov.gov.sg/

146. https://www.gosafeonline.sg/trustsg

147. Salmah Khairuddin, "Electronic Government in Malaysia", Malaysian Administrative Modernization And Management Planning Unit, Prime Minister's Department, Malaysia

148. http://www.mahadthai.com/html/index.html

149. Chumphol Krootkaew, "Thai e-Government today", 2nd Asian Document Style Standardization Symposium, 2004

150. Mr. Jirapon Tubtimhin, "Thailand e-Local Government Website Scorecard, A Tool to Create Digital Opportunity", WSIS Thematic Meeting, 2005

151. http://www.e-visavietnam.net/en/home.asp

152. http://www.comelec.gov.ph/

153. Al-Sobi, Faris, Vishanth Weerakkody and Shafi Al-Shafi (2009). European and Mediterranean Conference on Information Systems (12–13 April), Abu Dhabi, United Arab Emirates

154. http://www.ita.gov.om/ITAPortal/Government/Government.aspx

155. http://www.undp.org.lb/programme/governance/institutionbuilding/adminreform/egovernment/index.cfm

156. http://www.government.ae/web/guest/home/en

157. http://www.gov.il/firstgov/english

158. http://www.saudi.gov.sa/wps/portal/yesserRoot/home/!ut/p/b1/04_Sj9C Pykssy0xPLMnMz0vMAfGjzOId3Z2dgj1NjAz8zUMMDTxNzZ2NH U0NDd29DfWDU_P0_Tzyc1P1C7IdFQFV9YhO/dl4/d5/L2dBISEvZ0 FBIS9nQSEh/

159. http://portal.www.gov.qa/wps/portal/homepage

160. http://www.egov.gov.iq/egov-iraq/index.jsp?&lng=en
161. "e-Government actions in Europe Best European e-practices", e-Governance Academy, Tunisia
162. Busson, Alain and Alain Keravel (2005). Interoperable Government Providing Services: Key Questions and Solutions Analyzed through 40 Case Studies Collected in Europe. École des Hautes Études Commerciales de Paris
163. http://www.nhs.uk/Pages/HomePage.aspx
164. https://www.gov.uk/tell-us-once
165. http://wayf.dk/index.php
166. http://www.nemhandel.dk/#/forside
167. www.eesti.ee
168. http://www.digilugu.ee/portal/page/portal/Digilugu/ETerviseProjektid
169. http://www.ria.ee/26259
170. https://riigihanked.riik.ee/
171. http://archyvas.infobalt.lt/sl/2007/index_en.php?t=evaldzia
172. http://www.vrm.lt/go.php/lit/English
173. http://www.etenders.gov.ie/
174. http://taxpolicy.gov.ie/
175. http://www.revenue.ie/en/business/paye/guide/employers-guide-paye-ros.html
176. http://www.hse.ie/eng/services/find_a_service/bdm/certificates_ie
177. http://www.basis.ie/home/home.jsp?pcategory=10055&ecategory=10055&language=EN
178. Cap Gemini, S.A., and others (2009). Smarter, Faster, Better eGovernment. 8th Benchmark Measurement, November 2009. Prepared for European Commission Directorate General for Information Society and Media.
179. http://en.nav.gov.hu/e_services/E_returns_extra
180. https://orszaginfo.magyarorszag.hu/english
181. http://www.mvcr.cz/lstDoc.aspx?nid=5176&lang=cs
182. http://portal.justice.cz/justice2/uvod/uvod.aspx
183. http://cds.mfcr.cz/cps/rde/xchg/SID-3EA9846B-A1A5B3DC/cds/xsl/4230_7131.html?year=0
184. http://www.egov.bg/eGovPortal/appmanager/portal/portal
185. http://www.zdravenportal.bg/site/index.jsf
186. http://www.nap.bg/?lang=en
187. http://www.e-guvernare.ro/Default.aspx?LangID=4

188. Chatzidimitriou, Marios and Adamantios Koumpis (2008). Marketing One-stop E-Government Solutions: the European OneStopGov Project. IAENG International Journal of Computer Science, 35:1, IJCS_35_1_11.
189. http://en.wikipedia.org/wiki/Documento_Nacional_de_Identidad
190. http://e-uprava.gov.si/e-uprava/en/
191. Boštjan Tovornik, "A to Z of the Slovenian e-Government", European Conference of Public Administration
192. Angela Russo, "The MEPA - the Italian PA e-Marketplace", European e-government award, 2009
193. http://telematici.agenziaentrate.gov.it/Main/index.jsp
194. http://www.portaldocidadao.pt/PORTAL/pt
195. http://www.porbase.org/english/products/bnp.html
196. http://www.citius.mj.pt/portal/
197. http://www.english.umic.pt/index.php?option=com_content&task=view&id=44&Itemid=112
198. http://www.gov.mt/en/Pages/gov.mt%20homepage.aspx
199. https://www.gov.mt/en/myBills/Pages/myBillsMainPage.aspx
200. https://gov.mt/en/Services-And-Information/Business-Areas/Health%20Services/Pages/Health-Services-in-Malta.aspx
201. https://gov.mt/en/Services-And-Information/Business-Areas/Justice/Pages/Justice-in-Malta.aspx
202. https://www.mita.gov.mt/page.aspx?pageid=352
203. http://www.ermis.gov.gr/portal/page/portal/ermis/
204. http://www.gsis.gr/gsis_site/
205. http://www.syzefxis.gov.gr/Default.aspx?lang=2
206. http://www.grnet.gr/default.asp?pid=27&la=2
207. Deloitte (2010). User Expectations of a Life Events approach for Designing E-Government Services: Final Report prepared for the European Commission, DG Information Society and Media.
208. http://en.wikipedia.org/wiki/Carte_Vitale
209. http://www.impots.gouv.fr/portal/dgi/home?pageId=home&sfid=00
210. https://www.marches-publics.gouv.fr/?page=entreprise.EntrepriseHome
211. http://www.regierung.li/index.php?id=487
212. http://www.kmu.admin.ch/themen/00614/00715/01502/index.html?lang=en
213. https://www.simap.ch/shabforms/COMMON/application/applicationGrid.jsp?template=1&view=1&page=/MULTILANGUAGE/simap/content/start.jsp&language=EN

214. http://www.bmi.bund.de/EN/Themen/OeffentlDienstVerwaltung/
ModerneVerwaltung/DEMail/demail_node.html
215. http://www.115.de/nn_740530/EN/Project_D115/project_D115_node.
html?_nnn=true
216. http://www.guichet.public.lu/fr/index.html
217. http://www.eluxemburgensia.lu/R/RN=987133038&local_base=
SERIALS
218. http://www.anelo.lu/
219. http://eid.belgium.be/en/
220. http://www.ksz.fgov.be/
221. http://www.privacycommission.be/
222. Graham Hassall, "An Information Ecology Approach to Sustainable e-
GovernmentAmong Small Island Developing States in the Pacific",
http://www.academia.edu/1870876/An_Information_Ecology_
Approach_to_Sustainable_e-Government_Among_Small_Island_
Developing_States_in_the_Pacific
223. http://www.apsc.gov.au/about-the-apsc
224. http://archive.ict.govt.nz/plone/archive/about-egovt/listing_archives.
html
225. https://psi.govt.nz/home/default.aspx.
226. https://www.egov.gov.fj/default.aspx.
227. Official website of negp http://negp.gov.in/
228. http://deity.gov.in/content/mission-mode-projects,
229. Radha Chauhan, "National E-Governance Plan in India", UNU-IIST
Report No. 414
230. The e-Office Framework: A Way Forward for the Government, http://
darpg.nic.in/darpgwebsite_cms/Document/file/The_e-Office_
Framework.PDF
231. income tax, http://www.incometaxindia.gov.in/archive/e-brochure.pdf
232. http://www.irda.gov.in/Defaulthome.aspx?page=H1
233. Implementation of Mission Mode Project on 'Immigration, Visa and
Foreigner's Registration & Tracking (IVFRT)' http://www.pib.nic.in/
newsite/erelease.aspx?relid=61800
234. http://www.mca.gov.in/MCA21/dca/help/efiling/bulletin_banks.pdf
235. http://uidai.gov.in/
236. http://www.passportindia.gov.in/AppOnlineProject/welcomeLink
237. https://www.epostoffice.gov.in/informationTable.html
238. http://www.negp.gov.in/index.php?option=com_content&view=article
&id=157&Itemid=785

239. E banking and E governance, http://vivauniversity.files.wordpress.com/2012/09/mkgt303session5ebankingandgovernment.pdf

240. NeGP Guidelines for Operational Model for implementation of Mission Mode Projects by the Line Ministries/State Departments, http://mit.gov.in/content/mission-mode-projects

241. National Land Record Modernization Program – http://www.negp.gov.in/index.php?option=com_content&view=article&id=182&Itemid=809

242. National Transport Register, Informatics, Vol. 20, No. 2, Oct 2011.

243. Integrated Framework for Delivery of Services, e-District Mission Mode Project, Department of Electronics & Information Technology (DeitY), Ministry of Communications & Information Technology, Government of India, New Delhi

244. Report On Current State Assessment (As – Is Process Documentation), Commercial Taxes Mission Mode Project (CT MMP), Department of Revenue, Ministry of Finance, Government of India, New Delhi.

245. Piyush Gupta, R K Bagga, Sridevi Ayaluri, "Fostering e-Governance, Selected Compendium of Indian Initiatives", ICFAI Books, The ICFAI University Press.

246. e-Governance in Municipalities as part of JNNURM, http://www.nisg.org/knowledgecenter_docs/A05070001.pdf

247. Crime and Criminal Tracking Network and Systems (CCTNS) http://egovreach.in/uploads/presentation/ranchi/Crime_and_Criminal_Tracking_Network_&_Systems.pdf

248. "DACNET, an E-Governance Infrastructure for the Globalization of Indian Agriculture", Secretary, Central Insecticides Board & Registration Committee (CIBRC) Directorate of Plant Protection, NIC

249. E-Panchayat (Electronic Knowledge Based Panchayat), National Informatics Centre, Hyderabad.

250. NGP, Employment Exchange http://www.negp.gov.in/index.php?option=com_content&view=article&id=201&Itemid=828

251. Current Status of E-governance in Healthcare in the Large Hospitals, Vertika Shukla, M.Phil thesis, Christ University, Bangalore.

252. Ashok Kumar, "E-Governance in Education Sector", Gian Jyoti E-Journal, Volume 1, Issue 2

253. M.Vinayak Rao, A.N Siddiqui, Musharraf Sultan, "e-Public Distribution Monitoring system (e-PDMS)", NIC M.P State Centre, Bhopal.

254. NGP, Integrated Mission Mode Projects http://www.negp.gov.in/index.php?option=com_content&view=article&id=217&Itemid=844

255. Rajanish Das, Atashi Bhattacherjee, "Status of Common Service Center Program in India: Issues, Challenges and Emerging Practices for Rollout", Indian Institute of Management, Ahmadabad

256. e-Biz, Mission Mode Project and overview, http://www.dipp.gov.in/English/Ebiz/Integrated_MMP-eBiz.pdf

257. e-court, Department of Justice, Government of India, New Delhi, http://doj.gov.in/?q=node/39

258. "Implementation Guidelines For eProcurement rollout in states as a Mission Mode Project under National eGovernance Plan", Govt. of India, Ministry of Commerce and Industry, Department of Commerce.

259. Electronic Data Interchange (EDI) For Trade (eTrade), A NASSCOM ® initiative, http://egovreach.in/social/content/electronic-data-interchange-edi-trade-etrade

260. Neeta Verma, Alka Mishra, Sonal Kalra, National Portal of India*(http://india.gov.in), National Informatics Center

261. National Service Delivery Gateway , https://www.nsdg.gov.in/administration/aboutus.jsp

262. E-Government Databases: A Retrospective Study , http://www.ijcse.com/docs/IJCSE10-01-02-01.pdf

263. Per Myrseth, Jørgen Stang and Vibeke Dalberg, "A data quality framework applied to e-government metadata", http://www.semicolon.no/Myrseth_Shanghai_2011_paper.pdf

264. Data.gov Concept of Operations, http://www.data.gov/sites/default/files/attachments/data_gov_conops_v1.pdf

265. Dr.R.K.Mitra, Working Paper: Rise of E-Governance, http://cc.iift.ac.in/research/Docs/WP/13.pdf

266. Yola Georgiadou, Orlando Rodriguez-Pabón, and Kate Trinka Lance, "Spatial Data Infrastructure (SDI) and E-governance: A Quest For Appropriate Evaluation Approaches", http://downloads2.esri.com/campus/uploads/library/pdfs/119171.pdf.

267. Erhard Rahm, Hong Hai Do, "Data Cleaning: Problems and Current Approaches", http://wwwiti.cs.uni-magdeburg.de/iti_db/lehre/dw/paper/data_cleaning.pdf

268. Rajesh Chauhan & Amar Jeet Singh, "Generalizing Meta-Data Elements: Global Scenario and Indian Perspective", http://www.napsipag.org/pdf/5_1_article_2_chauhan___singh.pdf

269. Diakar Ray, Umesh Gulla , Shefali S Dash, "The Indian Approach to e-government Interoperability", http://joaag.com/uploads/6_1_-3_Ray_et_al.pdf

270. Dr. Sita Vanka, Mr. K. Sriram, Dr. Ashok Agarwal, "Critical Issues In E-Governance", http://www.csi-sigegov.org/critical_issues_on_e_governance.pdf

271. "Technical Standards for Interoperability Framework for E-Governance in India", Department of Information Technology, Ministry of Communications and Information Technology, Government of India, New Delhi

272. K.Meenakumari And P.Ambika, "Semantic Based Platform For Managing E-Governance Strategy Using Master Data Management Framework", http://www.ascent-journals.com/IJMRAE/Vol3No4/Paper-5.pdf

273. Sonali Agarwal, Neera Singh, Prof. G.N.Pandey, "implementation of Data Mining and Data Warehousing In E-Governance", http://www.ijcaonline.org/volume9/number4/pxc3871851.pdf

274. Dilip A. Joseph et. al. "Defining data administration and operational policies at the business objective level for e-governance application", power grid corporation of India limited.

275. Renu Budhiraja, "Challenges and Role of Standards in Building Interoperable e-Governance Solutions", http://www.csi-sigegov.org/E-Governence/e_Governance.pdf

276. "Focus On Informatics", http://elibrary.finec.ru/materials_files/3438486 11.pdf

277. Shailendra Singh, D. Singh Karaulia, "E-Governance: Information Security Issues", http://psrcentre.org/images/extraimages/1211468.pdf

278. Encryption and Digital Signatures , E-Government Manual, http://www.E-Government-handbuch.de"

279. Concept Note On Open Government Data, Moldova Rapid Governance Support Program, http://data.gov.md/wp-content/uploads/downloads/2011/04/Concept_Note_on_Open_Government_Data.pdf

280. Reform Measures And Policy Initiatives, http://commerce.nic.in/publications/OutBudget2011_12/Chapter_III.pdf

281. Amritesh, Subhas C. Misra, Jayanta Chatterjee "Examining Information Quality for e-Governance Services: Towards a Conceptual Model", http://www.ipcsit.com/vol31/023-ICIII2012-C20001.pdf

282. Dr. Mohammed T. Al-Sudairy and T. G. K Vasista, "SEMANTIC DATA INTEGRATION APPROACHES FOR E-GOVERNANCE" , http://airccse.org/journal/ijwest/papers/0111ijwest01.pdf

283. Debendra Kumar Mahalik, Outsourcing in e-Governance: A Multi Criteria Decision Making Approach, http://www.napsipag.org/pdf/5_1_article_3_mahalik.pdf

284. E-Governance Report Card, "Government of Bihar", http://southasia. oneworld.net/Files/e_governance_report_card.pdf

285. Per Myrseth, Jørgen Stang and Vibeke Dalberg, "A data quality framework applied to e-government metadata", http://www.semicolon.no/ Myrseth_Shanghai_2011_paper.pdf

286. Policy Document on "Identity and Access Management, Task Force on "Identity and Access Management", http://www.naavi.org/cl_editorial_ 07/negp_iam_draft1.pdf

287. Sameer Sachdeva, "White Paper on E-Governance Strategy in India", http://indiaegov.org/knowledgeexchg/egov_strategy.pdf

288. Sok Chuob, Manish Pokharel, Jong Sou Park, "The Future Data Center for E-Governance", ISBN: 978-1-4244-5427-3.

289. Enterprise Data Center Network Reference Architecture – Juniper Networks, the leader in high performance networking

290. A white paper on Cloud Computing for e-Governance. http://search.iiit. ac.in/uploads/CloudComputingForEGovernance.pdf.

291. Peter Mell and Tim Grance from National Institute of Standards and Technology, Information Technology Laboratory – a non-regulatory federal agency within the U.S. Department of Commerce – http://csrc.nist.gov/ groups/SNS/cloud-computing .

292. http://en.wikipedia.org/wiki/Cloud_computing.

293. http://www.networkcomputing.com/netdesign/cdmwdef.htm

294. "Cloud computing for e-government: Thailand experience" presented by Dr. Sak Seghoonthod ,President and CEO of Electronic government agency, Thialand.

295. http://www.ebizq.net/blogs/softwareinfrastructure/2009/08/the_seven_ elements_of_cloud_co.php

296. Cloud Controller – developed by Incontrinuum, the software development company introduced Cloud Controller to manage application designed and developed to automate the design, configuration, ordering, payment, chargeback, provisioning, maintenance, support, and reporting for cloud-based managed service and associated with SLAs.

297. Confederation of Indian Industry "Energy Efficiency Guidelines and Best Practices in Indian Datacenters" http://hightech.lbl.gov/dc-india/ documents/datacenterbook.pdf

298. "For appointment of an Agency for Establishment and Operations & Maintenance of physical cum IT infrastructure For Nagaland State Data Centre at Kohima"itngl.nic.in/it/SDC%20RFP%20Vol%20II-Nagaland. pdf RFP Volume 2.

299. http://www.datacentermap.com/

300. http://deity.gov.in/content/data-centre.

301. Dr. Kishori Lal Bansal, Sanjay Kumar Sharma, Satish Sood "Impact of Cloud Computing in Implementing Cost Effective E-governance Operations" , Gian Jyoti E-Journal, Volume 1, Issue 2 (Jan – Mar 2012), ISSN 2250-348X.

302. Oracle® Database Concepts 11g Release 1 (11.1), http://docs.oracle.com/cd/B28359_01/server.111/b28318/bus_intl.htm

303. Dr. Sonali Agarwal et. al. (2010) "Implementation of Data Mining and Data Warehousing In E-Governance". International Journal of Computer Applications 94:18–22, November 2010. Published By Foundation of Computer Science

304. Marijn Janssen (2006) "Enterprise Architecture Integration in E-Government" Proceedings of the 38th Hawaii International Conference on System Sciences – 2005 European and Mediterranean Conference on Information Systems (EMCIS) 2006, July 6–7 2006, Costa Blanca, Alicante, Spain.

305. Riley Thomas B.(2003) International Tracking Survey Report '03 Number Two "Knowledge Management and Technology" http://www.rileyis.com/publications/research_papers/tracking03/IntlTrackingRpt June03 no2.pdf.

306. Usman Muhammad Anwar, et al. (2006) "Multi-Agent Based Semantic E-Government Web Service Architecture" IEEE/WIC/ACM International Conferences on Web Intelligence and Intelligent Agent Technology - Workshops pp. 599–604.

307. Gregory B. White et al. (2006) "Introduction to the 2006 Minitrack on E-Government Security" Proceedings of the 39th Hawaii International Conference on System Sciences - 0-7695-2507-5/06/$20.00 (C) 2006 IEEE ieeexplore.ieee.org/iel5/10548/33364/ 01579445b.pdf?arnumber= 1579445.

308. Manuel Paul (2005) "A Model of E-Governance Based on Knowledge Management" Journal of Knowledge Management Practice, June 2005 http://i-policy.typepad.com/informationpolicy/2005/08/_a_model_of_ego.html

309. "Oracle Retail Extract Transform and Load" (n.d.) Retrieved August 2009, from www.oracle.com/applications/retail/bi/extract.html

310. "About Kiosk", (n.d.) E Governance of Government of West Bengal, Retrieved December 2009 www.wbgov.com/E-Gov/ENGLISH/Kiosk/AboutKiosk.asp.

311. Thomas Zwahr and Matthias Finger, (2004) "Enhancing the e-Governance model: Enterprise Architecture as a potential methodology to build a holistic framework" Proceedings of the International Conference on Politics and Information System: Technologies and Applications. Orlando, Florida, USA.

312. Saeed, M.et. al., (2010) "E-governance service delivery - an assessment of Community Information Centre Model in India" 2010 International Conference on Advances in ICT for Emerging Regions (ICTer), 978-1-4244-9041-7 11678055, 10.1109/ICTER.2010.5643274 Date of Current Version: 22 November 2010.

313. About Kiosk", (n.d.) E Governance of Government of West Bengal, Retrieved December 2009 www.wbgov.com/E-Gov/ENGLISH/Kiosk/AboutKiosk.asp.

314. Dolpanya, Kitsada et. al. (2009) "A Conceptual Framework for Investigating Suppliers' Participation in Business-to-Government (B2G) Electronic Auction Markets in the Thai Context". AMCIS 2009 Proceedings. Paper 678. http://aisel.aisnet.org/amcis2009/678.

315. Official Website of "Education for All" (n.d.) Retrieved on November 2010 from www.upefa.com/.

316. Official Website of National Polio Surveillances Project (NPSP) (n.d.) Retrieved on January 2011 http://www.npspindia.org/.

317. Cristianini, N., & Shawe-Taylor, J. (2000) An Introduction to Support Vector Machines. Cambridge University Press.

318. Suykens, J. A. K., & Vandewalle, J. (1999) "Least Squares Support Vector Machine Classifiers. Neural Processing Letters, 9, 293–300.

319. N.Chistianini and J. Shawe-Taylor, (2000) "An Introduction to Support Vector Machines, and other kernel-based learning methods", Cambridge University Press, 2000.

320. Data Preprocessing Techniques for Data Mining http://www.iasri.res.in/ebook/win_school_aa/notes/Data_Preprocessing.pdf

321. K R Muller and S Mika, (2001) "An Introduction to Kernel-Based Learning algorithms," IEEE Trans on Neural Networks, 2001, 122, pp.199–222.

322. Kernel Methods http://ttic.uchicago.edu/ďmcallester/ttic101-07/lectures/kernels/kernels.pdf

323. Suykens, J et. al, (2002) "Weighted Least Squares Support Vector Machines: Robustness and sparse approximation", Neurocomputing, 48:1–4, 85–105.

324. Weimin Huang, Leping Shen (2008) "Weighted Support Vector Regression Algorithm Based on Data Description", 2008 ISECS International

Colloquium on Computing, Communication, Control, and Management.

325. Navia-V´azquez et. al, (2001) "Weighted least squares training of support vector classifiers leading to compact and adaptive schemes". IEEE Transactions on Neural Networks, 12, 1047–1059.

326. Jiawei Han, Micheline Kamber (2006) "Data Mining: Concepts and Techniques, Third Edition (The Morgan Kaufmann Series in Data Management Systems)", Publisher: Morgan Kaufmann 2006.

327. Vapnik, V. (1998). "The support vector method of function estimation".

328. Support Vector Regression Machine http://ece.ut.ac.ir/classpages/F83/PatternRecognition/Papers/SupportVectorMachine/support-vector.pdf

329. Bezdek, J., (1981) "Pattern Recognition With Fuzzy Objective Function Algorithms", Plenum. New York.

330. "WEKA 3: Data Mining Software in Java" (n.d.) Retrieved March 2007 from http://www.cs.waikato.ac.nz/ml/weka/.

331. Mark Hall et al. (2009) "The WEKA Data Mining Software: An Update" SIGKDD Explorations, Volume 11, Issue 1.

332. "WEKA Data Mining Book" (n.d.) Retrieved March 2007 from http://www.cs.waikato.ac.nz/~ml/weka/book.html.

333. Merceron A. and Yacef K (2005) "Educational Data Mining: a Case Study", Proceedings of the 12th international Conference on Artificial Intelligence in Education AIED, pages 467–474. Amsterdam, The Netherlands, IOS Press, 2005.

334. Report on The system of education in India http://www.nokut.no/Documents/NOKUT/Artikkelbibliotek/Kunnskapsbasen/Konferanser/SU%20konferanser/Seminarer/Fagseminar_06/The%20System%20of%20Education%20in%20India.pdf

335. Official Website of District Information System for Education (DISE) (n.d.) Retrieved on November 2010 from www.dpepmis.org/.

336. "Worldwide Trends in the Human Development Index 1970–2010" (2010) UNDP official website Retrieved on January 2010 http://hdr.undp.org/en/data/trends/.

337. Kaur Harleen and Wasan Siri Krishan (2006) "Empirical Study on Applications of Data Mining Techniques in Healthcare" Journal of Computer Science 2 2: 194–200, 2006 ISSN 1549–3636 ©2006 Science Publications. www.scipub.org/fulltext/jcs/jcs22194-200.pdf .

338. "India Vision 2020" (2002) Planning Commission Report planningcommissionnic.in/reports/genrep/pl_vsn2020.pdf.

339. Data released by National Polio Surveillance Project: Govt. of India and WHO through internet URL: www.who.int/vaccines/casecount/afpextractnew.cfm.

340. Neera Singh, Sonali Agarwal, Ramesh Chandra Tripathi, "A Data Mining Perspective on the Prevalence of Polio in India" International Journal on Computer Science and Engineering (IJCSE) , ISSN : 0975-3397 Vol. 3 No. 2 Feb 2011, http://www.enggjournals.com/ijcse/doc/IJCSE11-03-02-166.pdf

341. A book "Data Warehousing: Concepts, Techniques, Products and Applications" by C.S.R.Prabhu.

342. Mirela DANUBIANU et.al "SOME ASPECTS OF DATA WAREHOUSING IN TOURISM INDUSTRY", The Annals of The "Stefan cel Mare" University Suceava. Fascicle of The Faculty of Economics and Public Administration, Volume 9, No.19, 2009.

343. Ruben Canlas Jr., "Governance, Human Development and Aid: Patterns Discovered Using Data Mining Techniques."

344. Jeffrey W. Seifert, "Data Mining and Homeland security: An overview", www.fas.org/sgp/crs/intel/RL31798.pdf.

345. William J. Krouse , "The Multi-State Anti-Terrorism Information Exchange (MATRIX) Pilot Project". http://www.fas.org/irp/crs/RL32536.pdf.

346. Sotarat Thammaboosadee and Udom Silparcha , "A framework for Criminal judicial reasoning system using data mining techniques", http://dlab.sit.kmutt.ac.th/event/InterConfer/Sotarat/Conf3IEEEDEST2008.pdf.

347. Justice XML structure Task Force, "The Justice XML Data Model:vOverview and status",http://it.ojp.gov/jxdm.

348. Stylios, G, Christodoulakis, D, Besharat, J, Vonitsanou, M, Kotrotsos, I, Koumpouri, A and Stamou, S. "Public Opinion Mining for Governmental Decisions" Electronic Journal of e-Government Volume 8 Issue 2 2010, (pp203–214), available online at www.ejeg.com.

349. Bethard, S.; Yu, Hong.,Thornton, Ashley., Hatzivassiloglou, Vasileios., and Jurafsky, Dan. 2004. "Automatic Ex-traction of Opinion Propositions and their Holders". (AAAI 2004),Springer.

350. Decman. Mitja, 2009. "Web 2.0 in eGovernment: The challenges and opportunities of Wiki in Legal Matters". Pro-ceedings of the 9th European Conference on eGovernment, pp 229–236.

351. Hian Chye Koh and Gerald Tan, "Data Mining Applications in Health-care", Journal of Healthcare Information Management — Vol. 19, No. 2.

352. Mary K. Obenshain, Mat, "Application Of Data Mining Techniques To Healthcare Data", Infection Control And Hospital Epidemiology, August 2004.

353. Rajanish Dass , "Data Mining in Banking and Finance: A Note for Bankers", Indian Institute of Management, Ahmadabad.

354. Vivek Bhambri, "Application of Data Mining in Banking Sector ", IJCSt Vol. 2, ISSue 2, June 2011, ISSN : 2229-4333(Print) |ISSN : 0976-8491(Online).

355. S. S. Kaptan, N S Chobey, "Indian Banking in Electronic Era", Sarup and Sons, Edition 2002.

356. S.P. Deshpande, Dr. V.M. Thakare, "Data Mining System And Applications: A Review".

357. Jiawei Han, Micheline Kamber (2006) "Data Mining: Concepts and Techniques, Third Edition (The Morgan Kaufmann Series in Data Management Systems)",Publisher: Morgan Kaufmann 2006.

358. Dorian Pyle, "Business Modeling and Data Mining", (The Morgan Kaufmann Series in Data Management Systems), ISBN-13: 978-1558606531

359. Sanjay Mohapatra, Mani Tiwari, "Using Business Intelligence for Automating Business Process in Insurance", International Journal of Advancements in Computing Technology Volume 1, Number 2, December 2009.

360. Alejandra Urtubiaa, J. Ricardo Pérez-Correaa, Alvaro Sotob, Philippo Pszczólkowski, "Using data mining techniques to predict industrial wine problem fermentations", Volume 18, Issue 12, December 2007, Pages 1512–1517.

361. Laila Mohamed ElFangary, Maryam Hazman, Alaa Eldin Abdallah Yassin, "Mining the Impact of Climate Change on Animal Production", International Journal of Computer Applications, Volume 59– No.18, December 2012.

362. Ahsan Abdullah, Stephen Brobst, Ijaz Pervaiz, Muhammad Umer, Azhar Nisar, "Learning Dynamics of Pesticide Abuse through Data Mining", Conferences in Research and Practice in Information Technology, Vol. 32.

363. Christopher J. Moran, Elisabeth N. Buicsiro, "Spatial data mining for enhanced soil map modeling", Int. J. Geographical Information Science, 2002, vol. 16, no. 6, 533–549

364. Luciana A. S. Romani, Ana Maria H. Ávila, Jurandir Zullo Jr.,Caetano Traina Jr., Agma J. M. Traina, "Mining Relevant and Extreme Patterns

on Climate Time Series with CLIPS Miner", Journal of Information and Data Management, Vol. 1, No. 2, June 2010, Pages 245–260.

365. Ryan S.J.d. Baker1, "Mining Data for Student Models", Department of Social Science and Policy Studies, Worcester Polytechnic Institute, 100, Institute Road, Worcester, MA 01609 USA

366. "Enhancing Teaching and Learning Through Educational Data Mining and Learning Analytics: An Issue Brief", Office of Educational Technology, U.S. Department of Education, USA

367. S.Anupama Kumar, Dr.Vijayalakshmi.M.N, "A Novel Approach in Data Mining Techniques for Educational Data", 2011 3rd International Conference on Machine Learning and Computing (ICMLC 2011).

368. Ryan S.J.d. Baker, Baker, R.S.J. "Data Mining for Education", Carnegie Mellon University, Pittsburgh, Pennsylvania, USA.

369. Vasile Paul Bresfelean, "Data Mining Applications in Higher Education and Academic Intelligence Management", Open Access Database, www.intechweb.org

370. Hogyeong Jeong, Gautam Biswas, "Mining Student Behavior Models in Learning-by-Teaching Environments", Department of Electrical Engineering and Computer Science, Vanderbilt University,

371. Sona Mardikyan, Bertan Badur, "Analyzing Teaching Performance of Instructors Using Data Mining Techniques", Informatics in Education, 2011, Vol. 10, No. 2, 245–257.

372. V.Ramesh, P.Parkavi, P.Yasodha, "Performance Analysis of Data Mining Techniques for Placement Chance Prediction", International Journal of Scientific & Engineering Research Volume 2, Issue 8, August-2011

373. Cristóbal Romero, Sebastián Ventura, Enrique García, "Data mining in course management systems: Model case study and tutorial", Department of Computer Sciences and Numerical Analisys, University of Córdoba

374. Gary M. Weiss, " Data Mining in the Telecommunications Industry".

375. Chapman, A. (2007). SWOT analysis. Retrieved October 10, 2007, from http://www.businessballs.com/swotanalysisfreetemplate.htm

376. JRC European Commission. (2007). SWOT (strengths weaknesses opportunities and threats) analysis. Retrieved October 19, 2007, from http://forlearn.jrc.es/guide/2_design/meth_swot-analysis.htm

377. http://pestleanalysis.com/

378. http://www.kantakji.com/fiqh/files/env/ty2.pdf

379. http://riccentre.ca/wp-content/uploads/2012/01/Session-1-Takeaways Guidlines.pdf

Index

E Governance Data Center, Data Warehousing and Data Mining: Vision to Realities, 271–274.